D0722104

A Panorama of Harmonic Analysis

© 1999 by
The Mathematical Association of America (Incorporated)
Library of Congress Catalog Card Number 99-62756

Complete Set ISBN 0-88385-000-1
Vol. 26 ISBN 0-88385-031-1

Printed in the United States of America

Current Printing (last digit):
10 9 8 7 6 5 4 3 2 1

The Carus Mathematical Monographs

Number Twenty-Seven

A Panaroma of Harmonic Analysis

Steven G. Krantz
Washington University in St. Louis

Published and Distributed by
THE MATHEMATICAL ASSOCIATION OF AMERICA

THE
CARUS MATHEMATICAL MONOGRAPHS

Published by
THE MATHEMATICAL ASSOCIATION OF AMERICA

———

Committee on Publications
William Watkins, *Chair*

Carus Mathematical Monographs Editorial Board
Steven G. Krantz, *Editor*
Robert Burckel
John B. Conway
Giuliana P. Davidoff
Gerald B. Folland
Leonard Gillman

The following Monographs have been published:

MAA Service Center
P. O. Box 91112
Washington, DC 20090-1112
800-331-1MAA   FAX: 301-206-9789

In homage to
Abram Samoilovitch Besicovitch (1891–1970),
an extraordinary geometric analyst.

Contents

Preface

The history of modern harmonic analysis dates back to eighteenth century studies of the wave equation. Explicit solutions of the problem

$$\frac{\partial^2 u}{\partial x^2} - \frac{\partial^2 u}{\partial t^2} = 0$$
$$u(x, 0) = \sin jx \quad \text{or} \quad \cos jx$$

were constructed by separation of variables. The question naturally arose whether an arbitrary initial data function $f(x)$ could be realized as a superposition of functions $\sin jx$ and $\cos jx$. And thus Fourier analysis was born.

Indeed it was Fourier [FOU] who, in 1821, gave an explicit means for calculating the coefficients a_j and b_j in a formal expansion

$$f(x) \sim \sum_j a_j \cos jx + \sum_j b_j \sin jx.$$

The succeeding 150 years saw a blossoming of Fourier analysis into a powerful set of tools in applied partial differential equations, mathematical physics, engineering, and pure mathematics. Fourier analysis in the noncompact setting—the Fourier transform—was developed, and the Poisson summation formula was used, to pass back and forth between Fourier series and the Fourier transform.

The 1930s and 1940s were a relatively quiet time for Fourier analysis but, beginning in the 1950s, the focus of Fourier analysis became *singular integrals*. To wit, it rapidly developed that, just as the Hilbert transform is the heart of the matter in the study of Fourier series of one variable, so singular integrals usually lie at the heart of any nontrivial problem of several-variable linear harmonic analysis. The Calderón-Zygmund theory of singular integrals blossomed into the Fefferman-Stein-Weiss theory of Hardy spaces; Hardy spaces became the focus of Fourier analysis.

In the 1980s, two seminal events served to refocus Fourier analysis. One was the David-Journé-Semmes $T(1)$ theorem on the L^2 boundedness of (not necessarily translation-invariant) singular integrals. Thus an entirely new perspective was gained on which types of singular integrals could induce bounded operators. Calderón commutators, Hankel operators, and other classical objects were easy pickings using the powerful new tools provided by the $T(1)$ theorem, and more generally the $T(b)$ theorem.

The other major event of the 1980s was the development of wavelets by Yves Meyer in 1985. Like any good idea, wavelet theory has caused us to "reinvent" Fourier analysis. Now we are no longer bound to model every problem on sine waves and cosine waves. Instead, we can invent a Fourier analysis to suit any given problem. We have powerful techniques for localizing the problem both in the space variable and the phase variable. Signal processing, image compression, and many other areas of applied mathematics have been revolutionized because of wavelet theory.

The purpose of the present book is to give the uninitiated reader an historical overview of the subject of Fourier analysis as we have just described it. While this book is considerably more polished than merely a set of lectures, it will use several devices of the lecture: to prove a theorem by considering just an example; to explain an idea by considering only a special case; to strive for clarity by not stating the optimal form of a theorem.

We shall not attempt to explore the more modern theory of Fourier analysis of locally compact abelian groups (i.e., the theory of group characters), nor shall we consider the Fourier analysis of non-abelian groups (i.e., the theory of group representations). Instead, we shall restrict attention to the classical Fourier analysis of Euclidean space.

Prerequisites are few. We begin with a quick and dirty treatment of the needed measure theory and functional analysis. The rest of the book uses only elementary ideas from undergraduate real analysis. When a sophisticated idea is needed, it is quickly introduced in context.

It is hoped that the reader of this book will be imbued with a sense of how the subject of Fourier analysis has developed and where it is heading. He will gain a feeling for the techniques that are involved and the applications of the ideas. Even those with primary interests in other parts of mathematics should come away with a knowledge of which parts of Fourier analysis may be useful in their discipline, and also where to turn for future reading.

In this book we indulge in the custom, now quite common in harmonic analysis and partial differential equations, of using the same letters (often C or C' or K) to denote different constants—even from line to line in the same proof. The reader unfamiliar with this custom may experience momentary discomfort, but will soon realize that the practice streamlines proofs and increases understanding.

It is a pleasure to thank the many friends and colleagues who have read and commented on portions of various preliminary drafts of this book. I mention particularly Lynn Apfel, Brian Blank, John McCarthy, Dylan Retsek, Richard Rochberg, Mitchell Taibleson, Guido Weiss, and Steven Zemyan. J. Marshall Ash and Victor Shapiro gave me expert help with the history and substance of the theory of multiple Fourier series.

And now for some special thanks: Robert Burckel aimed his sharp and critical eye at both my English and my mathematics; the result is a greatly improved manuscript. Jim Walker contributed many incisive remarks, particularly stemming from his expertise in wavelets; in addition, Walker provided figures (that were generated with his software FAWAV) for Chapter 7. Gerald B. Folland kept me honest and, acting as Chair of the Carus Committee, helped me to craft the mathematics and to ensure that this project came out as it should. The assistance provided by these three scholars has been so extraordinary, and so extensive, that I sometimes feel as though this manuscript has four authors.

Of course responsibility for the extant manuscript lies entirely with me. I am always happy to receive reports of errors or suggestions for improvement.

Steven G. Krantz
St. Louis, Missouri

Overview of Measure Theory and Functional Analysis

0.1 Pre-Basics

We take it for granted that the reader of this book has some acquaintance with elementary real analysis. The book [KRA1], or Rudin's classic [RUD1], will provide ample review.

However, the reader may be less acquainted with measure theory. The review that we shall now provide will give even the complete neophyte an intuitive understanding of the concept of measure, so that the remainder of the book may be appreciated. In fact, the reading of the present book will provide the reader with considerable motivation for learning more about measure theory.

By the same token, we shall follow the review of measure theory with a quick overview of some elementary functional analysis. Again, the reader already conversant with these ideas may consider our treatment to be a review. The beginner will gain a sufficient appreciation from this outline to be able to proceed through the rest of the book. The applications of functional analysis presented in this book—to questions about convergence of Fourier series, for instance—should serve as motivation for the reader to learn more about functional analysis.

0.2   A Whirlwind Review of Measure Theory

The motivation for measure theory on the real line $\mathbb{R}$ is a desire to be able to measure the length of any set. At first, we only know that the length of an interval $[a, b]$ (or $[a, b)$ or $(a, b]$ or (a, b)) is $b - a$. If the set S is the union of finitely many disjoint intervals, then we may say that the length of S is the sum of the lengths of these component intervals. But what if S is more complicated—say a Cantor set, or the complement of a Cantor set?

It turns out that, as long as the Axiom of Choice is present in our set theory, then it is impossible to assign a length to every subset of $\mathbb{R}$ consistent with certain elementary properties that one expects a notion of "length" to satisfy. We shall not provide the relevant proof of this assertion (see [FOL1] for the details), but use this fact for motivation of the notion of "measurable set." A measurable set is one that we are allowed to measure. We shall, in this section, give a brief overview of the measure theory of Lebesgue (sometimes called Lebesgue-Tonelli measure theory).

Now let us look at these matters from another point of view. Let $S \subseteq \mathbb{R}$ be any subset. If $U \supset S$ is any open set, then of course we may write

$$U = \bigcup_{j=1}^{\infty} I_j,$$

where each I_j is an open interval and the I_j are pairwise disjoint. And if $I_j = (a_j, b_j)$ then we may set $m(I_j) = b_j - a_j$. Finally we define $m(U) = \sum_j m(I_j)$ and

$$m(S) = \inf_{U \supset S} m(U),$$

where the infimum is taken over all open sets U that contain S. [Note here that, out of homage to tradition, we use the letter m to denote the length, or measure, of a set.] The mapping m, as defined here, is what is called an "outer measure."

In simple examples, where one has some intuition as to what the length of a set S ought to be, this new definition gives the expected answer. For example, if S is an interval (open, closed, or half-open), then $m(S)$ is easily seen to be the ordinary length obtained by subtracting the endpoints. If S is a countable set, $S = \{s_j\}_{j=1}^{\infty}$, then for any $\epsilon > 0$ we may set

$$U_\epsilon = \bigcup_{j=1}^{\infty} (s_j - \epsilon/2^j, s_j + \epsilon/2^j).$$

Then $m(U_\epsilon) \le 2\epsilon$. Since $U_\epsilon \supset S$, and since $\epsilon > 0$ is arbitrary, we may conclude that $m(S) = 0$. As an exercise, the reader may wish to calculate that the length of the Cantor "middle thirds" set is 0.

We have already said that there are certain elementary properties that one expects a notion of "length" to possess. These include

(0.2.1) $$m\left(\bigcup_{j=1}^{\infty} S_j\right) = \sum_{j=1}^{\infty} m(S_j)$$

when the sets S_j are pairwise disjoint and

(0.2.2) $$m\left(\bigcup_{j=1}^{\infty} S_j\right) \le \sum_{j=1}^{\infty} m(S_j)$$

when the sets S_j are not necessarily pairwise disjoint. In fact, Property (0.2.1) may *fail* unless the sets S_j are restricted to be "measurable." Now let us say what the technical condition of measurability actually means.

It is convenient to specify the measurable sets indirectly (the class we are describing here has the technical name of "σ-algebra"). We certainly want open sets to be measurable, and we want the collection of measurable sets to be closed under complementation, countable unions, and countable intersections (the collection of sets generated in this fashion is called the "Borel sets"). We also want any set with measure zero to be measurable, since (as will be seen below) sets of measure zero are the omnipresent "fudge factors" of the subject. Thus the collection of measurable sets will be the smallest collection of sub-sets of $\mathbb{R}$ that contains the open sets, contains the sets of measure zero, and has the closure properties just described.

Measurable sets have the following appealing properties (we omit the rather technical proofs, but see [FOL1] for details):

(0.2.3) $m(A \cup B) \le m(A) + m(B)$;

(0.2.4) If $A \subseteq B$ then $m(A) \le m(B)$;

(0.2.5) If $A \subseteq B$ then $m(A) + m(B \setminus A) = m(B)$;

(0.2.6) $m\left(\bigcup_{j=1}^{\infty} A_j\right) \le \sum_{j=1}^{\infty} m(A_j)$;

(0.2.7) If the sets A_j are pairwise disjoint, then

$$m\left(\bigcup_{j=1}^{\infty} A_j\right) = \sum_{j=1}^{\infty} m(A_j);$$

(0.2.8) If $A_1 \subseteq A_2 \subseteq \cdots$ and $\cup A_j = A$ then $\lim_{j \to \infty} m(A_j) = m(A)$;

(0.2.9) If $A_1 \supseteq A_2 \supseteq \cdots$ and $\cap A_j = A$ and $m(A_1) < \infty$ then $\lim_{j \to \infty} m(A_j) = m(A)$.

The main points to remember about measurable sets are these: **(i)** open sets are measurable; **(ii)** closed sets are measurable; **(iii)** count-able unions and intersections of measurable sets are measurable; **(iv)** the complement of a measurable set is measurable; **(v)** any set of measure zero is measurable.

An important property of Lebesgue measure is "regularity." Let S be a measurable set that has finite measure. Then there is a compact set K and an open set U such that $K \subseteq S \subseteq U$ and

$$m(S \setminus K) < \epsilon \quad \text{and} \quad m(U \setminus S) < \epsilon.$$

In Appendix I, for instance, you will see arguments that demonstrate the usefulness of inner approximation by compact sets ("inner regular-ity") and outer approximation by open sets ("outer regularity").

In measure theory it is common to make statements like

Property $P(x)$ holds almost everywhere.

This statement means that there is a set E of measure zero with the property that $P(x)$ holds for all x in $\mathbb{R} \setminus E$. With practice, you will learn to read such a statement as saying that P holds at all but a very small set of exceptional points.

Now let us speak of measurable functions. The "atom," or unit, of measurable functions is the characteristic (or indicator) function: if $S \subseteq \mathbb{R}$ is a measurable set then we set

$$\chi_S(x) = \begin{cases} 1 & \text{if} \quad x \in S \\ 0 & \text{if} \quad x \in \mathbb{R} \setminus S. \end{cases}$$

We call χ_S the *characteristic function* of the set S. A *simple function* is any finite linear combination of characteristic functions; thus a typical simple function has the form

$$s = \sum_{j=1}^{K} a_j \chi_{S_j}$$

for measurable sets $S_1, \ldots, S_K$ and complex numbers $a_1, \ldots, a_K$. A *measurable function* f is any function that is the pointwise limit of a sequence $\{s_j\}$ of simple functions. If $f \geq 0$, then it can always be assumed that $0 \leq s_1 \leq s_2 \leq \cdots \leq f$ and $s_j \rightarrow f$. [A more direct, and traditional, definition of measurable function is that it is a function $f : \mathbb{R} \rightarrow \mathbb{C}$ with the property that $f^{-1}(U)$ is a measurable set whenever $U \subseteq \mathbb{C}$ is open. In particular, continuous functions are certainly measurable. This definition is equivalent to the one just given.][1]

We allow a measurable function to assume the values $+\infty$ and $-\infty$. It follows that countable families $\{f_j\}_{j=1}^{\infty}$ of measurable functions are closed under sup, inf, lim sup, lim inf, and lim. Thus measurable functions are suitable units for real analysis.

---

[1] Properly speaking, the study of measurable functions is the study of *equivalence classes* of functions. For we identify two measurable functions when they differ only on a set of measure zero. In this text we follow the custom of working analysts and do not usually observe this distinction.

Next we shall define the integral. If $s = \sum_{j=1}^{K} a_j \chi_{S_j}$ is a simple function, then we define

$$\int s(x)\, dx = \sum_{j=1}^{K} a_j m(S_j).$$

With this definition, it is clear that if s, t are simple functions and if $s(x) \le t(x)$ for all x, then $\int s(x)\, dx \le \int t(x)\, dx$.

If f is a nonnegative, measurable function and if s_j is a sequence of nonnegative simple functions such that $s_j \nearrow f$, then the sequence $\{\int s_j(x)\, dx\}$ is monotone increasing and we define

$$\int f(x)\, dx \equiv \lim_{j \to \infty} \int s_j(x)\, dx,$$

whether this limit is finite or infinite. If f is a real-valued, measurable function, then set

$$f^{+} = f \cdot \chi_{\{x : f(x) > 0\}}$$

and

$$f^{-} = |f| \cdot \chi_{\{x : f(x) < 0\}}.$$

It is not difficult to see that f^{+}, f^{-} are measurable. We set

$$\int f(x)\, dx \equiv \int f^{+}(x)\, dx - \int f^{-}(x)\, dx,$$

provided one of the summands on the right is finite. Finally, if u and v are real-valued functions and $f(x) = u(x) + iv(x)$ is complex-valued, then we define

$$\int f(x)\, dx = \int u(x)\, dx + i \int v(x)\, dx,$$

provided *both* summands are finite. Thus we have defined the Lebesgue integral. Notice that it is almost immediate from the definition that

$$\left| \int f(x)\, dx \right| \le \int |f(x)|\, dx.$$

A measurable function f with the property that $\int |f(x)|\,dx$ exists and is finite is called "integrable." In case $\int |f(x)|^p\,dx$ exists and is finite, $0 < p < \infty$, then we say that f is "p^{th}-power integrable." The set of pth-power integrable functions forms a linear space and is denoted by L^p (in honor of Henri Lebesgue (1875–1941)).

It is interesting to note that the Riemann integral is formed by breaking up the *domain* of the integrand (the function being integrated), while the Lebesgue integral is formed by breaking up the *range* of the integrand. The primary advantage of the Lebesgue integral over the Riemann integral is that the Lebesgue integral comprises a broader class of limiting operations and of functions that we can integrate. The standard theorem in Riemann integration theory is that $\int f_j(x)\,dx \rightarrow \int f(x)\,dx$ provided that $f_j \rightarrow f$ *uniformly* on the domain of integration. The condition of uniform convergence is too restrictive for subtle analysis. In the Lebesgue theory, we have three classic convergence theorems that serve in most circumstances. We shall now state (but not prove) them. Refer to [FOL1] for the details.

The Lebesgue Monotone Convergence Theorem:
Suppose that $0 \le f_1 \le f_2 \le \cdots$ *are measurable functions. Let* $f(x) = \lim_{j\to\infty} f_j(x)$. *Then*

$$\lim_{j\to\infty} \int f_j(x)\,dx = \int f(x)\,dx.$$

The Lebesgue Dominated Convergence Theorem:
Let $\{f_j\}$ *be measurable functions on* $\mathbb{R}$. *Assume that there is a measurable function* $g \ge 0$ *such that* $|f_j| \le g$ *for all* j *and* $\int g(x)\,dx < \infty$. *If* $f_j(x) \rightarrow f(x) \in \mathbb{C}$ *for almost every* x, *then*

$$\lim_{j\to\infty} \int f_j(x)\,dx = \int f(x)\,dx.$$

Fatou's Lemma:
If $f_j \ge 0$ *are measurable functions, then*

$$\liminf_{j\to\infty} \int f_j(x)\,dx \ge \int \liminf_{j\to\infty} f_j(x)\,dx.$$

These three fundamental theorems are sometimes referred to as **LMCT**, **LDCT**, and **FL**, respectively. Although we shall not prove them, we note that each of these theorems is (nearly) logically equivalent to the other two.

We end this section with a compendium of useful facts from measure theory. Again see [FOL1] for a discursive discussion of each.

Proposition 0.2.10 (Fubini's Theorem) *Let f be a continuous, bounded function on $I \times J$, where I and J are (possibly infinite) intervals in the real line. Then*

$$\int_I \int_J f(x, y) \, dx dy = \int_J \int_I f(x, y) \, dy dx.$$

Fubini's theorem is actually true for "product-measurable functions," but a careful consideration of this concept would take us far afield. A thumbnail version of Fubini's theorem is that if f is integrable in one order of the iterated integrals, then it is integrable in any other order and the integrations yield the same answer. When we apply Fubini's theorem in the sequel, the reader should not worry about the technical hypotheses of the theorem. The full details of Fubini's theorem may be found in [FOL1] or [RUD2].

When properly developed, Fubini's theorem has the consequence that we can perform M-dimensional Lebesgue integration by simply iterating 1-dimensional integrals, just as we do in calculus with the Riemann integral.

One useful consequence of Fubini's theorem is this: If a measurable function f on $\mathbb{R}^2$ satisfies $\int_{\mathbb{R}^2} |f(x)| \, dx < \infty$, then the functions $f_{x_1}(t) \equiv f(x_1, t)$ are in $L^1(\mathbb{R})$ for almost every choice of x_1; likewise, the functions $f^{x_2}(t) \equiv f(t, x_2)$ are in $L^1(\mathbb{R})$ for almost every choice of x_2. [Of course, there are analogous statements for p^{th}-power integrability.] We shall apply this observation decisively in Chapter 3.

Let p be a number with $1 \leq p \leq \infty$. When p is used in the context of an L^p space, then the number p is called an *index*. The *conjugate index* (or conjugate exponent) to p is the unique number p' that satisfies

$$\frac{1}{p} + \frac{1}{p'} = 1.$$

The custom here is that, if $p = 1$, then $p' = \infty$; and if $p' = 1$, then $p = \infty$.

Proposition 0.2.11 (Hölder) *Let p and p' be conjugate exponents with $1 \leq p, p' < \infty$. If $f \in L^p$ and $g \in L^{p'}$, then $f \cdot g \in L^1$ and*

$$\left| \int f(x) \cdot g(x) \, dx \right| \leq \left[\int |f(x)|^p \, dx \right]^{1/p} \cdot \left[\int |g(x)|^{p'} \, dx \right]^{1/p'}.$$

[**Remark:** When $p = p' = 2$, this inequality is known as the Cauchy-Schwarz, or sometimes the Cauchy-Schwarz-Bunjakowski, inequality.]

Proposition 0.2.12 (Jensen's Inequality) *Let $[a, b] \subseteq \mathbb{R}$ be an interval of finite length. Let f be an integrable function on $[a, b]$. Let ϕ be a convex function on $\mathbb{R}$. Then*

$$\phi \left[\frac{1}{b-a} \int_a^b f(x) \, dx \right] \leq \frac{1}{b-a} \int_a^b \phi \circ f(x) \, dx.$$

Sketch of Proof. If ϕ is linear, then the asserted inequality is in fact an equality. If instead ϕ is a general convex function, then ϕ is the upper envelope of its tangent lines (Figure 1). Thus taking the supremum over the equalities for the tangent lines gives the result. □

Proposition 0.2.13 (Minkowski) *Let $1 \leq p < \infty$. If f and g are L^p functions on $[a, b]$, then*

$$\left[\int_a^b |f(x) + g(x)|^p \, dx \right]^{1/p}$$
$$\leq \left[\int_a^b |f(x)|^p \, dx \right]^{1/p} + \left[\int_a^b |g(x)|^p \, dx \right]^{1/p}.$$

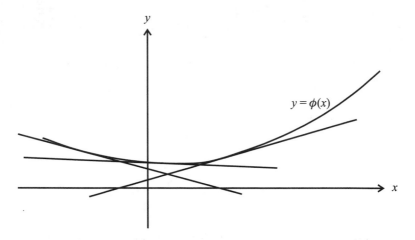

Figure 1.   A convex function is the upper envelope of its tangent lines.

Of course the integral is a continuous analogue of a sum (as the theory of the Stieltjes integral makes explicit). Thus it is no surprise that Minkowski's inequality has the following generalization:

Proposition 0.2.14 (Minkowski's Integral Inequality) *Let f be a measurable function on* $\mathbb{R}^N \times \mathbb{R}^M$. *Let* $1 \le p < \infty$. *Then*

$$\left[\int \left| \int f(x, y) \, dx \right|^p dy \right]^{1/p} \le \int \left[\int |f(x, y)|^p \, dy \right]^{1/p} dx.$$

Remark: Notice that, if we accept that $\|f\|_{L^p} \equiv [\int |f(x)|^p \, dx]^{1/p}$ is a norm (see below), then Minkowski's integral inequality says that "the norm of the sum is less than or equal to the sum of the norms."

The next theorem is a paradigm for theorems in harmonic analysis. While it appears to be an innocent theorem about differentiation of integrals (and it is almost intuitively obvious when f is continuous), the method of proof and the insights provided by this theorem prove to be fundamental in the modern view of the subject. We shall be studying this theorem in a more general context, and proving it, later in the book (Theorem 6.2.4).

Theorem 0.2.15 *Let f be a function on $\mathbb{R}^N$ that is integrable on compact sets. Then, for almost every $x \in \mathbb{R}^N$,*

$$\lim_{R \to 0} \frac{1}{m(B(x, R))} \int_{B(x,R)} f(t)\, dt = f(x).$$

The following cognate result is important in practice.

Theorem 0.2.16 *Let f be a function on $\mathbb{R}^N$ that is integrable on compact sets. Then, for almost every $x \in \mathbb{R}^N$,*

$$\lim_{R \to 0} \frac{1}{m(B(x, R))} \int_{B(x,R)} |f(t) - f(x)|\, dt = 0.$$

Points x that satisfy this conclusion are called Lebesgue points.

As a final note about Lebesgue measure, we record the Chebyshev inequality.

Lemma 0.2.17 (Chebyshev's Inequality) *Let $f \in L^p(\mathbb{R}^N)$. Then for any $\lambda > 0$ we have*

$$m\{x \in \mathbb{R}^N : |f(x)| > \lambda\} \le \frac{\int |f(x)|^p\, dx}{\lambda^p}.$$

Proof. We calculate that

$$m\{x \in \mathbb{R}^N : |f(x)| > \lambda\} \le \frac{1}{\lambda^p} \int_{\{x:|f(x)|>\lambda\}} |f(x)|^p\, dx$$

$$\le \frac{1}{\lambda^p} \int_{\mathbb{R}^N} |f(x)|^p\, dx \qquad \square$$

When f, g are L^1 functions on $\mathbb{R}^N$, we define their *convolution* to be

$$f * g(x) = \int_{\mathbb{R}^N} f(x - t)g(t)\, dt.$$

It follows from Fubini's theorem that $f * g$ is finite almost everywhere and, indeed, $f * g$ is an L^1 function. Moreover,

$$f * g(x) = \int_{\mathbb{R}^N} f(x - t)g(t)\, dt = \int_{\mathbb{R}^N} f(t)g(x - t)\, dt.$$

We refer the reader to [FOL1] or [RUD2] for the details. Convolutions are part and parcel of Fourier analysis, and arise naturally in almost every problem that we shall consider.

When we do analysis on $\mathbb{T} = \mathbb{R}/2\pi\mathbb{Z}$ (as when we study Fourier series), then the model for our integration theory will be the interval $[0, 2\pi)$ (see the next section for details about $\mathbb{T} = \mathbb{R}/2\pi\mathbb{Z}$). In that context, it is convenient to define the convolution of two L^1 functions f and g by

$$f * g(x) = \frac{1}{2\pi} \int_0^{2\pi} f(x - t)g(t)\, dt = \frac{1}{2\pi} \int_0^{2\pi} f(t)g(x - t)\, dt.$$

The prefactor of $1/2\pi$ is very natural in view of the formulas we shall be deriving.

We conclude this discussion of Lebesgue measure with a few words about abstract measure theory. The setup is that we are given a set X, a σ-algebra $\mathcal{A}$ of subsets of X (i.e., a collection of sets that is closed under the operations of countable union, countable intersection, and complementation), and a scalar-valued function μ on $\mathcal{A}$. Based on these primitives, one can build a "measure theory" analogous to the standard Lebesgue theory for Euclidean space. All of the customary facts about measure theory—including LMCT, LDCT, FL, Hölder's inequality, Minkowski's inequality, and so forth will continue to hold in this more general setting.

In a rigorous development of measure theory (see [FOL1]), one constructs a σ-algebra on $\mathbb{R}^N$ by using as generators the inverse images of measurable sets in $\mathbb{R}$ under the coordinate projection mappings. Having done so, one shows that for an integrable function of N variables, the integral may be evaluated by integrating in each variable successively, and in any order. This is part of the content of Fubini's theorem.

One example of an abstract measure is the so-called *Dirac measure* on the σ-algebra $\mathcal{A}$ of Borel sets in $\mathbb{R}^N$. This is the set-function

$$\delta_0 : \mathcal{A} \to \mathbb{R}$$

that is given by $\delta_0(S) = 1$ if $0 \in S$ and $\delta(S) = 0$ if $0 \notin S$. This measure is commonly thought of as the "point mass at the origin."

Another common method for constructing measures (the so-called "absolutely continuous" measures) is the following. Let ϕ be an integrable, nonnegative function on $\mathbb{R}^N$—with respect to the usual Lebesgue measure. Define, for S a Borel set,

$$\mu(S) = \int_S \phi(x)\, dx.$$

This is the measure with weight ϕ.

Next we turn to a cursory treatment of basic functional analysis.

0.3   The Elements of Banach Space Theory

Let X be a finite-dimensional vector space over $\mathbb{R}$ and let $\{\mathbf{x}^1, \ldots, \mathbf{x}^k\}$ be a basis for X. Let $X \ni x = \sum \alpha_j \mathbf{x}^j$ and define the norm of x to be

$$\|x\| = \sum_j |\alpha_j| \tag{0.3.1}$$

The mapping

$$X \ni x = \sum_{j=1}^k \alpha_j \mathbf{x}^j \leftrightarrow (\alpha_1, \ldots, \alpha_k) \in \mathbb{R}^k$$

shows that X is isomorphic to $\mathbb{R}^k$, both topologically and algebraically. In particular, X is complete.

A norm on a real or complex vector space X is a nonnegative, real-valued function $\| \ \|$ on the space that satisfies

(0.3.2) $\|\lambda x\| = |\lambda| \cdot \|x\|$ for any scalar λ and any element $x \in X$;

(0.3.3) $\|x + y\| \leq \|x\| + \|y\|$ for any $x, y \in X$.

It is worth noting that any norm on a finite-dimensional vector space is equivalent to any other. For a proof, let $\| \ \|_*$ be any norm on $\mathbb{R}^k$. We will show that it is equivalent to the norm $\| \ \|$ defined in (0.3.1) above. Let M denote the maximum of the expressions $\|\mathbf{x}^j\|_*$, $j = 1, \ldots, k$. Notice that if $x = \alpha_1 \mathbf{x}^1 + \cdots \alpha_k \mathbf{x}^k$, then

$$\|x\|_* = \|\alpha_1 \mathbf{x}^1 + \cdots \alpha_k \mathbf{x}^k\|_*$$

$$\leq \sum_{j=1}^{k} |\alpha_j| \|\mathbf{x}^j\|_*$$

$$\leq M \cdot \sum_{j=1}^{k} |\alpha_j|$$

$$= M \cdot \|x\|.$$

That is half of what we want to prove.

For the reverse inequality, let $S \subseteq \mathbb{R}^k$ be the set of points with $\| \ \|$-norm equal to 1. Of course S is compact in the topology induced by $\| \ \|$. The preceding inequality then shows that $\| \ \|_*$ is $\| \ \|$-continuous. Hence S is compact in the $\| \ \|_*$-topology. Observe that $\| \ \|_*$ is a continuous and positive function on S, hence has a positive minimum value m. It follows that $\|x\|_* \geq m$ for $x \in S$; hence

$$\|x\|_* \geq m \cdot \|x\|$$

for any $x \in \mathbb{R}^k$. That is what we wished to prove.

Matters are different for infinite-dimensional spaces. In harmonic analysis, we wish to study spaces of functions, and these are almost never finite-dimensional. Thus we must come to grips with infinite-dimensional spaces. A (complex) normed linear space that is *complete* in the topology induced by the norm is called a *Banach space*. We shall now discuss the particular Banach spaces that will be most important for our studies in the present book. For further reading, consult [RUD2], [RUD3], [FOL1], [DUS].

Example 0.3.4 For $1 \leq p < \infty$ we define $L^p(\mathbb{R})$ to be the linear space of those measurable functions f such that $\int |f(x)|^p \, dx < \infty$. The norm on L^p is

$$\|f\|_{L^p} \equiv \|f\|_p \equiv \left[\int |f(x)|^p \, dx \right]^{1/p}.$$

When $p = \infty$ we define $L^\infty(\mathbb{R})$ to be the linear space of functions that are "essentially bounded." Thus $f \in L^\infty(\mathbb{R})$ if

$$\mathcal{M} = \sup\{M : m\{x : |f(x)| > M\} > 0\} \tag{0.3.4.1}$$

is finite. [Observe that (0.3.4.1) is just a measure-theoretic version of boundedness. The definition says, in effect, that $|f|$ is bounded by $\mathcal{M}$ except on a set of measure 0, and that $\mathcal{M}$ is the least number with this property.] We let $\mathcal{M}$ be the L^∞ norm of f, which we denote by $\|f\|_{L^\infty} \equiv \|f\|_\infty$. It is common to refer to the L^1 functions as the "integrable" functions and to L^∞ as the "essentially bounded" functions.

It is not obvious, but it is true, that each L^p space is a Banach space. In other words, if $\{f_j\}$ is a Cauchy sequence in the L^p topology, then there is a function $f \in L^p$ such that $\|f_j - f\|_p \to 0$.

On occasion, we shall refer to L^p_{loc}, $1 \leq p < \infty$. Here we say that $f \in L^p_{\text{loc}}$ if, whenever ϕ is a continuous function with compact support, then ϕf lies in L^p (in other words, f is p^{th}-power integrable on compact sets).

The L^p norms, $1 \leq p \leq \infty$, are useful tools because of the following essential properties (which repeat in new language some of the properties from our measure theory review):

Minkowski's Inequality:

$$\|f + g\|_p \leq \|f\|_p + \|g\|_p.$$

Minkowski's Integral Inequality:

$$\left\|\int f(x,\,\cdot)\,dx\right\|_{L^p} \le \int \|f(x,\,\cdot)\|_{L^p}\,dx.$$

Hölder's Inequality:

$$\left|\int f(x)\cdot g(x)\,dx\right| \le \|f\|_p\cdot\|g\|_q,$$

provided that $(1/p) + (1/q) = 1$ *(the pair $p = 1, q = \infty$ is allowed here) and $f \in L^p$, $g \in L^q$.*

We shall not prove these inequalities, but refer the reader to [FOL1] and [RUD2] for these and all matters concerning measure theory. ☐

We do not consider L^p for $p < 1$ because those spaces are not locally convex and do not satisfy analogues of either Minkowski's or Hölder's inequalities. Note that each L^p is clearly infinite-dimensional, since the functions $f_j = \chi_{[j,j+1]}$, $j = 1, 2, \ldots$ are linearly independent.

Note in passing that a linear subspace Y of a Banach space X is called a *Banach subspace* of X if Y is complete in the norm inherited from X. As an example, if X is the space of bounded, continuous functions on the real line—equipped with the supremum norm—then the space Y of bounded, continuous functions that vanish at the origin is a Banach subspace.

The fundamental objects in Banach space theory are the bounded linear operators. A linear operator

$$L : X \to Y,$$

with X and Y Banach spaces, is said to be *bounded* if there is a finite constant $M > 0$ such that

$$\|Lx\|_Y \le M\|x\|_X \tag{0.3.5}$$

for every $x \in X$. Clearly a bounded linear operator is continuous in the topological sense, and the converse is true as well (exercise). We define the *operator norm* of L to be the infimum of all constants M that satisfy (0.3.5); the operator norm is denoted by $\|L\|_{\text{op}}$. As an exercise, the reader may check that this operator norm possesses all the usual properties of a norm.

We will be especially interested in the bounded linear operators from a Banach space X to the space $\mathbb{C}$ of complex numbers. Such a linear operator $L : X \to \mathbb{C}$ is called a *bounded linear functional* on X. The space of bounded linear functionals on X, equipped with the operator norm, is itself a Banach space and is called the *dual space* of X; it is denoted by X^*. See the further discussion below (Definition 0.3.8).

Like measure theory, elementary Banach space theory has three fundamental results. We shall now state, but not prove, them.

The Open Mapping Principle:

Let $L : X \to Y$ be a bounded, surjective linear mapping of Banach spaces. Then L is an open mapping (i.e., L takes open sets to open sets). In particular, if L is one-to-one, then there is a constant $K > 0$ such that

$$\|Lx\|_Y \geq K \|x\|_X$$

for all $x \in X$.

The Uniform Boundedness Principle (Banach-Steinhaus Theorem):

Let $L_j : X \to Y$ be bounded linear mappings of Banach spaces. Then one and only one of the following two alternatives holds:

(0.3.6) *There is a finite constant M such that*

$$\|L_j\|_{\text{op}} \leq M, \qquad \text{for all } j.$$

(0.3.7) *There is a dense set E in X such that*

$$\sup_{j} \|L_j e\| = +\infty \qquad \text{for all } e \in E.$$

The Hahn-Banach Theorem:

Let X be a Banach space and $Y \subset X$ a Banach subspace. Let $L : Y \to \mathbb{C}$ be a bounded linear functional with norm M. Then there is a bounded linear functional $\widetilde{L} : X \to \mathbb{C}$ such that $\widetilde{L}(y) = L(y)$ whenever $y \in Y$, and the operator norm of $\widetilde{L}$ equals M.

Remark: In fact the analogue of the Hahn-Banach theorem for linear operators $L : X \to Y$ between Banach spaces is false. The reader may wish to explore this matter as an exercise, or see [DUS].

The theory of Fourier series and integrals will provide us with ample illustrations of the use of these important Banach space theorems. We note for now that, in our applications to Fourier series, the Banach spaces in question will usually be $L^p(\mathbb{T})$ (see the beginning of Section 1.1 for more on these spaces). Here the formal definition of $\mathbb{T}$ is

$$\mathbb{T} = \mathbb{R}/2\pi\mathbb{Z}.$$

Of course we define these function spaces just as on the real line, with integration taking place instead on $[0, 2\pi)$. It is customary to write a function on $\mathbb{T}$ either as $f(e^{it})$ or as $f(t)$, $0 \leq t < 2\pi$. While we shall usually choose the former, the reader should become comfortable with either notation.

We conclude this section with a brief treatment of the notion of dual space.

Definition 0.3.8 Let X be a Banach space. Then X^* denotes the space of all bounded linear functionals on X. That is, $\alpha \in X^*$ if and only if $\alpha : X \to \mathbb{C}$ is a linear operator such that $|\alpha x| \leq C\|x\|$ for all $x \in X$ and some constant $0 \leq C < \infty$. The least constant C for which this inequality holds is called the *norm* of α. We call X^* the *dual space*, or the *dual* of X.

The space X^* is a vector space over the same field of scalars over which X is defined. If X is complete then so is X^* (in the topology induced by the norm defined in the last paragraph).

It is worth noting that X is "contained in" $(X^*)^* \equiv X^{**}$ in a natural way. Indeed, we may define the mapping

$$X \rightarrow X^{**}$$

$$x \mapsto \widehat{x},$$

where $\widehat{x}(\phi) = \phi(x)$ whenever $\phi \in X^*$. This mapping is one-to-one but is not always onto. In the special instance when the mapping *is* onto we say that the space X is *reflexive*.

The characterization of the duals of the L^p spaces is a fundamental result of measure theory. We record, but do not prove, the result now:

Theorem 0.3.9 *Let* $1 \leq p < \infty$. *Then the dual of* $L^p(\mathbb{R}^N)$ *is* $L^{p'}(\mathbb{R}^N)$, *where* $1/p + 1/p' = 1$. *More precisely, if* $\alpha \in (L^p(\mathbb{R}^N))^*$, *then there exists a function* $g \in L^{p'}(\mathbb{R}^N)$ *such that*

$$\alpha(f) = \int_{\mathbb{R}^N} f(x) \cdot g(x)\, dx$$

for all $f \in L^p(\mathbb{R}^N)$. *Conversely, if* $g \in L^{p'}$, *then the mapping*

$$f \longmapsto \int f \cdot g$$

defines a bounded linear functional on L^p. *In fact the norm of the functional is equal to* $\|g\|_{L^{p'}}$.

It is an interesting fact that L^1 is not the dual of *any* Banach space, and it seems to be an intractable problem to calculate the dual of $L^\infty(\mathbb{R}^N)$. Calculating the dual of a given Banach space is often quite a difficult task.

Using the Hahn-Banach theorem, it is possible to prove the following useful characterization of the L^p norm of a function $f \in L^p$,

$1 \leq p \leq \infty$:

$$\|f\|_{L^p} = \sup_{\|g\|_{L^{p'}}=1} \left| \int f(x) \cdot g(x)\, dx \right|.$$

Let X be a Banach space and $A \subseteq X$. We say that A is *dense* in X if for each $x \in X$ there is a sequence $\{a_j\} \subseteq A$ such that $a_j \to x$ in the Banach space topology. For instance, X could be $L^p(\mathbb{R}^N)$, $1 \leq p < \infty$, and A could be the linear space of continuous functions with compact support.

Definition 0.3.10 Let X, Y be Banach spaces and let $L : X \to Y$ be a bounded linear operator. That is, there is a finite constant $C > 0$ such that $\|Lx\|_Y \leq C \cdot \|x\|_X$ for all $x \in X$. Then the *adjoint* operator L^* of L, a linear operator from Y^* to X^*, is defined by the equation

$$L^*(\beta) = \beta \circ L$$

or

$$[L^*(\beta)][x] = \beta[Lx] \qquad (0.3.10.1)$$

when $\beta \in Y^*$. Notice that $L^*(\beta)$ is well-defined by (0.3.10.1) and that $L^*(\beta)$ is thereby an element of X^* when $\beta \in Y^*$.

The following is an elementary application of the definitions and propositions presented thus far:

Proposition 0.3.11 *Let X, Y be Banach spaces. Let $L : X \to Y$ be a bounded linear operator. Then L has dense range if and only if L^* is injective. And if L^* has dense range then L is injective. The converse of this last statement (i.e., if L is injective then L^* has dense range) is true provided that X is reflexive.*

Proof. Suppose that L has dense range. Now suppose that $y^* \in Y^*$ satisfies $L^* y^* = 0$. Then, for any $x \in X$,

$$0 = [L^* y^*](x) = y^*(Lx).$$

Since L has dense range, this says that y^* annihilates all of Y. Therefore y^* is the zero functional, that is, $y^* = 0$. Thus L^* is injective. That proves one fourth of the proposition. The other parts are proved similarly. For an example to illustrate the qualification in the last statement, consult [RUD3].   □

Remark: The phrase "dense range" in the statement of the proposition cannot be replaced with "surjective." As an example, consider the operator $L : L^2(\mathbb{R}) \to L^2(\mathbb{R})$ given by $L(f)(x) = x \cdot f(x)$. Then L is one-to-one but $L^* = L$ is not surjective.

0.4   Hilbert Space

We now consider a space whose norm comes from an inner product.

Definition 0.4.1 If V is a vector space over $\mathbb{C}$, then a function

$$\langle \cdot, \cdot \rangle : V \times V \to \mathbb{C}$$

is called a *positive-definite Hermitian inner product* if

(0.4.1.1) $\langle v_1 + v_2, w \rangle = \langle v_1, w \rangle + \langle v_2, w \rangle$,   all $v_1, v_2, w \in V$;

(0.4.1.2) $\langle \alpha v, w \rangle = \alpha \langle v, w \rangle$,   all $\alpha \in \mathbb{C}, v, w, \in V$;

(0.4.1.3) $\langle v, w \rangle = \overline{\langle w, v \rangle}$,   all $v, w \in V$;

(0.4.1.4) $\langle v, v \rangle \geq 0$,   all $v \in V$;

(0.4.1.5) $\langle v, v \rangle = 0$ iff $v = 0$.

Here are some simple examples of complex vector spaces equipped with positive-definite Hermitian inner products:

Example 0.4.2 Let $V = \mathbb{C}^n$ with its usual vector space structure. For $v = (v_1, \ldots, v_n), w = (w_1, \ldots, w_n)$ elements of $\mathbb{C}^n$ define

$$\langle v, w \rangle = \sum_{j=1}^{n} v_j \overline{w}_j.$$

The verification of properties (0.4.1.1)–(0.4.1.5) is immediate.   □

Example 0.4.3 Let V be the set of all continuous, complex-valued functions on $[0, 1]$, with addition and scalar multiplication defined in the usual way. For $f, g \in V$ define

$$\langle f, g \rangle = \int_0^1 f(x) \overline{g(x)} \, dx.$$

The verification of properties (0.4.1.1)–(0.4.1.5) is routine.   □

If V is a complex vector space and $\langle \cdot, \cdot \rangle$ a positive-definite Hermitian inner product on V, then we define an associated norm by

$$\|v\| = \sqrt{\langle v, v \rangle}$$

(recall, for motivation, the relation between norm and inner product on $\mathbb{R}^N$). Fundamental to the utility of this norm is that it satisfies the *Cauchy-Schwarz* inequality:

$$|\langle v, w \rangle| \le \|v\| \cdot \|w\|.$$

An immediate consequence of the Cauchy-Schwarz inequality is the triangle inequality: If $v, w \in V$ then

$$\|v + w\| \le \|v\| + \|w\|.$$

[**Exercise:** Confirm the triangle inequality by squaring both sides, using the connection between the norm and the inner product, and invoking Cauchy-Schwarz.]

A complex vector space V with positive-definite Hermitian inner product $\langle \cdot, \cdot \rangle$, which is complete under the associated norm $\| \ \|$, is called a *Hilbert space*.

Example 0.4.4 As in Example 0.4.2, let

$$V = \mathbb{C}^n \quad , \quad \langle v, w \rangle = \sum_{j=1}^{n} v_j \overline{w}_j \quad , \quad \|v\| = \sqrt{\sum_{j=1}^{n} |v_j|^2}.$$

This norm provides the usual notion of distance (or metric)

$$d(v, w) = \|v - w\| = \sqrt{\sum_{j=1}^{n} |v_j - w_j|^2}.$$

We know that $\mathbb{C}^n$, equipped with this metric, is complete. □

Example 0.4.5 Let $V = L^2(\mathbb{R}^N)$ and let the inner product be

$$\langle f, g \rangle = \int_{\mathbb{R}^N} f(x) \overline{g(x)} \, dx$$

(notice that Hölder's inequality with $p = p' = 2$ guarantees that the integral on the right makes sense). Then the norm on V is

$$\|f\|_{L^2} \equiv \left(\int_{\mathbb{R}^N} |f(x)|^2 \, dx \right)^{1/2}.$$

It is an important fact that this particular inner product space is complete in the norm just described. This is the quintessential example of a Hilbert space. □

A subset $A \subseteq H$ is called a *(Hilbert) subspace* of H if A is a vector subspace of H *and* A is a Hilbert space under the inner product inherited from H.

Let H be a Hilbert space. If $x, y \in H$ then we say that x is *orthogonal* to y and write $x \perp y$ if $\langle x, y \rangle = 0$. If $x \in H$ and $S \subseteq H$, then we write $x \perp S$ if $x \perp s$ for all $s \in S$, and we say that x is orthogonal to S. We write $S^\perp$ for the set of all such x, and we call $S^\perp$ the *annihilator* of S. Notice that $S^\perp$ will always be a closed subspace—that is, a Hilbert subspace—of H.

The Parallelogram Law:
If $x, y \in H$ then

$$\|x + y\|^2 + \|x - y\|^2 = 2\|x\|^2 + 2\|y\|^2$$

(the sum of the squares of the diagonals of a parallelogram equals the sum of the squares of the four sides—see Figure 1).

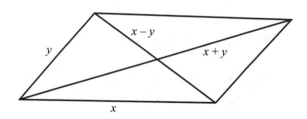

Figure 1.   The parallelogram law.

If $K \subseteq H$ is a closed subspace, then every $x \in H$ can be written in a unique way as $x = x_1 + x_2$ where $x_1 \in K, x_2 \in K^\perp$. The element x_2 may be characterized as the unique element of smallest norm in $x + K = \{x + k : k \in K\}$. If $x \in K$ then $x_1 = x, x_2 = 0$. The element x_1 is called the (orthogonal) *projection* of x onto K. The map $x \mapsto x_1$ is called the (orthogonal) *projection* of H onto K.

Fix an element $x_0 \in H$. Then the function $\phi : x \mapsto \langle x, x_0 \rangle$ is a continuous linear functional. The most fundamental result in elementary Hilbert space theory is that every continuous linear functional on H has this form.

The Riesz Representation Theorem:
If $\phi : H \to \mathbb{C}$ is a continuous linear functional, then there is a unique $p \in H$ such that

$$\phi(x) = \langle x, p \rangle \quad \text{for all} \quad x \in H. \tag{0.4.6}$$

A consequence of Riesz's theorem is this: If $\phi : H \to \mathbb{C}$ is a continuous linear functional and $K = \{x : \phi(x) = 0\}$, then $K^\perp$ has dimension 1.

A subset $\{u_\alpha\}_{\alpha \in A}$ is called *orthonormal* if each u_α has norm 1 and, in addition, $u_\alpha \perp u_\beta$ whenever $\alpha \neq \beta$. We say that the orthonormal set (or system) $\{u_\alpha\}_{\alpha \in A}$ is *complete* if the assumption $v \perp u_\alpha$ for all α implies that $v = 0$. [The underlying meaning here is that no nonzero vector can be added to the orthonormal system $\{u_\alpha\}_{\alpha \in A}$ to produce a larger one.] A complete orthonormal system is called an *orthonormal basis* for the Hilbert space, and this terminology is justified by Theorem 0.4.9 below.

In this book, all the Hilbert spaces that we shall consider will be separable (i.e., will possess a countable dense subset). As a result, each Hilbert space will have a *countable orthonormal basis*. In practice, we will produce such a basis explicitly.

The main fact about a complete orthonormal system is that it plays essentially the same role as a basis does in the finite-dimensional theory. This is the idea of Theorem 0.4.9 below. As a preliminary, here is a result that extends a standard idea of Euclidean geometry (dropping a perpendicular from a point to a line or plane) to the infinite-dimensional situation.

Proposition 0.4.7 *Let $u_1, \ldots, u_k \in H$ be orthonormal and let $x \in H$. We set $\alpha_j = \langle x, u_j \rangle$ for each j. The nearest element to x in the space U spanned by $u_1, \ldots, u_k$ is $u = \sum_{j=1}^k \alpha_j u_j$. The distance from x to u is*

$$\|x - u\| = \sqrt{\|x\|^2 - \sum_{j=1}^k |\alpha_j|^2}.$$

Corollary 0.4.8 (Bessel's Inequality) *With notation as in the proposition,*

$$\sum_{j=1}^k |\alpha_j|^2 \leq \|x\|^2.$$

Theorem 0.4.9 (Bessel, Riesz, Fischer) *Let $\{u_j\} \subseteq H$ be a complete orthonormal system. If $x \in H$ and $\alpha_j = \langle x, u_j \rangle$ for each j, then the partial sums $\sum_{j=1}^{N} \alpha_j u_j$ converge in the Hilbert space norm to x as $N \to \infty$. Furthermore, $\|x\|^2 = \sum_{j=1}^{\infty} |\alpha_j|^2$.*

Conversely, if $\sum |\beta_j|^2 < \infty$, then there is an $x \in H$ such that $\langle x, u_j \rangle = \beta_j$ for all j. For this x we have $\|x\|^2 = \sum |\beta_j|^2$ and $x = \sum_{j=1}^{\infty} \beta_j u_j$ (in the customary sense of $x = \lim_{N \to \infty} \sum_{j=1}^{N} \beta_j u_j$ in the Hilbert space topology).

Example 0.4.10 Let ℓ^2 be the set of complex sequences $\alpha = \{a_j\}_{j=1}^{\infty}$ such that $\sum_{j=1}^{\infty} |a_j|^2 < \infty$. When $\alpha = \{a_j\}_{j=1}^{\infty}, \beta = \{b_j\}_{j=1}^{\infty}$ are elements of ℓ^2, we define $\langle \alpha, \beta \rangle = \sum_{j=1}^{\infty} a_j \overline{b}_j$. Then ℓ^2 is a complex vector space with a positive-definite Hermitian inner product. If H is now any Hilbert space with a countable complete orthonormal system $\{u_j\}$, then the map

$$\Phi : H \ni x \mapsto \{\langle x, u_j \rangle\}_{j=1}^{\infty} \in \ell^2$$

is (by Theorem 0.4.9) linear, one-to-one, onto, and continuous with a continuous inverse. In fact Φ is norm-preserving. Thus, in particular, ℓ^2 is a Hilbert space. (In this sense it is the canonical Hilbert space having a countable, complete orthonormal system.) $\qquad\square$

Corollary 0.4.11 *If $\beta = \{b_j\}_{j=1}^{\infty}$ is an element of ℓ^2, then*

$$\|\beta\|_{\ell^2} = \left(\sum_{j=1}^{\infty} |b_j|^2 \right)^{1/2} = \sup_{\substack{\alpha \in \ell^2 \\ \|\alpha\| \leq 1}} |\langle \alpha, \beta \rangle|.$$

0.5   Two Fundamental Principles of Functional Analysis

In this section we formulate and prove two principles of functional analysis that will serve us well in the sequel. The proofs are at the end of the section.

Functional Analysis Principle I (FAPI)

Let X be a Banach space and S a dense subset. Let $T_j : X \to X$ be linear operators. Suppose that

(0.5.1) For each $s \in S$, $\lim_{j\to\infty} T_j s$ exists;

(0.5.2) There is a finite constant $C > 0$, independent of x, such that

$$\|T_j x\|_X \le C \cdot \|x\|$$

for all $x \in X$ and all indices j.

Then $\lim_{j\to\infty} T_j x$ exists for every $x \in X$.

For the second Functional Analysis Principle, we need a new notion. If $T_j : L^p \to L^p$ are linear operators on L^p (of some Euclidean space), then we let

$$T^* f(x) \equiv \sup_j |T_j f(x)|.$$

We call T^* the *maximal function* associated to the T_j.

Functional Analysis Principle II (FAPII)

Let $1 \le p < \infty$ and suppose that $T_j : L^p \to L^p$ are linear operators (these could be L^p on the circle, or the line, or $\mathbb{R}^N$, or some other Euclidean setting). Let $S \subseteq L^p$ be a dense subset. Assume that

(0.5.3) For each $s \in S$, $\lim_{j\to\infty} T_j s(x)$ exists in $\mathbb{C}$ for almost every x;

(0.5.4) There is a universal constant $0 \le C < \infty$ such that, for each $\alpha > 0$,

$$m\{x : T^* f(x) > \alpha\} \le \frac{C}{\alpha^p} \|f\|_{L^p}^p.$$

[Here m denotes Lebesgue measure.]

Then, for each $f \in L^p$,

$$\lim_{j \to \infty} T_j f(x)$$

exists for almost every x.

The inequality hypothesized in Condition (0.5.4) is called a *weak-type (p, p) inequality* for the maximal operator T^*. Weak-type inequalities have been fundamental tools in harmonic analysis ever since M. Riesz's proof of the boundedness of the Hilbert transform (see Section 1.6 and the treatment of singular integrals in Chapter 6). A classical L^p estimate of the form

$$\|Tf\|_{L^p} \leq C \cdot \|f\|_{L^p}$$

is sometimes called a *strong-type* estimate.

We shall use FAPI (Functional Analysis Principle I) primarily as a tool to prove norm convergence of Fourier series and other Fourier-analytic entities. We shall use FAPII (Functional Analysis Principle II) primarily as a tool to prove pointwise convergence of Fourier series and other Fourier entities. We devote the remainder of this section to proofs of these two basic principles.

Proof of FAPI. Let $f \in X$ and suppose that $\epsilon > 0$. There is an element $s \in \mathcal{S}$ such that $\|f - s\| < \epsilon/3(C + 1)$. Now select $J > 0$ so large that if $j, k \geq J$ then $\|T_j s - T_k s\| < \epsilon/3$. We calculate, for such j, k, that

$$\|T_j f - T_k f\| \leq \|T_j f - T_j s\| + \|T_j s - T_k s\| + \|T_k s - T_k f\|$$

$$\leq \|T_j\|_{\mathrm{op}} \|f - s\| + \frac{\epsilon}{3} + \|T_k\|_{\mathrm{op}} \|f - s\|$$

$$\leq C \cdot \|f - s\| + \frac{\epsilon}{3} + C \cdot \|f - s\|$$

$$< \frac{\epsilon}{3} + \frac{\epsilon}{3} + \frac{\epsilon}{3}$$

$$= \epsilon.$$

This establishes the result. Note that the converse holds by the Uniform Boundedness Principle. ☐

Proof of FAPII. This proof parallels that of FAPI, but it is more technical.

Let $f \in L^p$ and suppose that $\delta > 0$ is given. Then there is an element $s \in S$ such that $\|f - s\|_{L^p}^p < \delta$. Let us assume for simplicity that f and the $T_j f$ are real-valued (the complex-valued case then follows from linearity). Fix $\epsilon > 0$ (independent of δ). Then

$$m\{x : \left|\limsup_{j\to\infty} T_j f(x) - \liminf_{j\to\infty} T_j f(x)\right| > \epsilon\}$$

$$\leq m\{x : \left|\limsup_{j\to\infty}[T_j(f - s)](x)\right| > \epsilon/3\}$$

$$+ m\{x : \left|\limsup_{j\to\infty}(T_j s)(x) - \liminf_{j\to\infty}(T_j s)(x)\right| > \epsilon/3\}$$

$$+ m\{x : \left|\limsup_{j\to\infty}[T_j(s - f)](x)\right| > \epsilon/3\}$$

$$\leq m\{x : \sup_j \left|[T_j(f - s)](x)\right| > \epsilon/3\}$$

$$+ 0$$

$$+ m\{x : \sup_j \left|[T_j(s - f)](x)\right| > \epsilon/3\}$$

$$= m\{x : T^*(f - s)(x) > \epsilon/3\}$$

$$+ 0$$

$$+ \{x : T^*(s - f)(x) > \epsilon/3\}$$

$$\leq C \cdot \frac{\|f - s\|_{L^p}^p}{[\epsilon/3]^p} + 0 + C \cdot \frac{\|f - s\|_{L^p}^p}{[\epsilon/3]^p}$$

$$< \frac{2C\delta}{\epsilon/3}.$$

Since this estimate holds no matter how small δ, we conclude that

$$m\{x : \left|\limsup_{j\to\infty} T_j f(x) - \liminf_{j\to\infty} T_j f(x)\right| > \epsilon\} = 0.$$

This completes the proof of FAPII, for it shows that the desired limit exists almost everywhere. $\qquad\square$

CHAPTER 1

Fourier Series Basics

1.0 The Pre-History of Fourier Analysis

In the middle of the eighteenth century much attention was given to the problem of determining the mathematical laws governing the motion of a vibrating string with fixed endpoints at 0 and π (Figure 1). An elementary analysis of tension shows that if $y(x, t)$ denotes the ordinate of the string at time t above the point x, then $y(x, t)$ satisfies the *wave equation*

$$\frac{\partial^2 y}{\partial t^2} = a^2 \frac{\partial^2 y}{\partial x^2}.$$

Here a is a parameter that depends on the tension of the string. A change of scale will allow us to assume that $a = 1$. [For completeness, we include a derivation of the wave equation at the end of this section.]

Figure 1. A vibrating string.

In 1747 d'Alembert showed that solutions of this equation have the form

$$y(x, t) = \frac{1}{2} [f(t + x) + g(t - x)], \qquad (1.0.1)$$

where f and g are "any" functions of one variable. [The following technicality must be noted: the functions f and g are initially specified on the interval $[0, \pi]$. We extend f and g to $[-\pi, 0]$ and to $[\pi, 2\pi]$ by odd reflection. Continue f and g to the rest of the real line so that they are 2π-periodic.]

In fact the wave equation, when placed in a "well-posed" setting, comes equipped with two boundary conditions:

$$y(x, 0) = \phi(x)$$

$$\partial_t y(x, 0) = \psi(x).$$

If (1.0.1) is to be a solution of this boundary value problem, then f and g must satisfy

$$\frac{1}{2} [f(x) + g(-x)] = \phi(x) \qquad (1.0.2)$$

and

$$\frac{1}{2} [f'(x) + g'(-x)] = \psi(x). \qquad (1.0.3)$$

Integration of (1.0.3) gives a formula for $f(x) - g(-x)$. That and (1.0.2) give a system that may be solved for f and g with elementary algebra.

The converse statement holds as well: For any functions f and g, a function y of the form (1.0.1) satisfies the wave equation (Exercise). The work of d'Alembert brought to the fore a controversy which had been implicit in the work of Daniel Bernoulli, Leonhard Euler, and others: what is a "function"? [We recommend the article [LUZ] for an authoritative discussion of the controversies that grew out of classical studies of the wave equation. See also [LAN].]

It is clear, for instance, in Euler's writings that he did not perceive a function to be an arbitrary "rule" that assigns points of the domain to points of the range; in particular, Euler did not think that a function could be specified in a fairly arbitrary fashion at different points of the domain. Once a function was specified on some small interval, Euler thought that it could only be extended in one way to a larger interval (was he thinking of real-analytic functions?). Therefore, on physical grounds, Euler objected to d'Alembert's work. He claimed that the initial position of the vibrating string could be specified by several different functions pieced together continuously, so that a single f could not generate the motion of the string.

Daniel Bernoulli solved the wave equation by a different method (separation of variables) and was able to show that there are infinitely many solutions of the wave equation having the form

$$\phi_j(x, t) = \sin jx \cos jt.$$

Proceeding formally, he posited that all solutions of the wave equation satisfying $y(0, t) = y(\pi, t) = 0$ and $\partial_t y(x, 0) = 0$ will have the form

$$y = \sum_{j=1}^{\infty} a_j \sin jx \cos jt.$$

Setting $t = 0$ indicates that the initial form of the string is $f(x) \equiv \sum_{j=1}^{\infty} a_j \sin jx$. In d'Alembert's language, the initial form of the string is $\frac{1}{2}(f(x) - f(-x))$, for we know that

$$0 \equiv y(0, t) = f(t) + g(t)$$

(because the endpoints of the string are held stationary), hence $g(t) = -f(t)$. If we suppose that d'Alembert's function is odd (as is $\sin jx$, each j), then the initial position is given by $f(x)$. Thus the problem of reconciling Bernoulli's solution to d'Alembert's reduces to the question of whether an "arbitrary" function f on $[0, \pi]$ may be written in the form $\sum_{j=1}^{\infty} a_j \sin jx$.

Since most mathematicians contemporary with Bernoulli believed that properties such as continuity, differentiability, and periodicity were preserved under (even infinite) addition, the consensus was that arbitrary f could *not* be represented as a (even infinite) trigonometric sum. The controversy extended over some years and was fueled by further discoveries (such as Lagrange's technique for interpolation by trigonometric polynomials) and more speculations.

In the 1820s, the problem of representation of an "arbitrary" function by trigonometric series was given a satisfactory answer as a result of two events. First there is the sequence of papers by Joseph Fourier culminating with the tract [FOU]. Fourier gave a formal method of expanding an "arbitrary" function f into a trigonometric series. He computed some partial sums for some sample f's and verified that they gave very good approximations to f. Secondly, Dirichlet proved the first theorem giving sufficient (and very general) conditions for the Fourier series of a function f to converge pointwise to f. *Dirichlet was one of the first, in 1828, to formalize the notions of partial sum and convergence of a series*; his ideas certainly had antecedents in work of Gauss and Cauchy.

For all practical purposes, these events mark the beginning of the mathematical theory of Fourier series (see [LAN]).

Fourier's Point of View

In [FOU], Fourier considered variants of the following basic question. Let there be given an insulated, homogeneous rod of length π with initial temperature at each $x \in [0, \pi]$ given by a function $f(x)$ (Figure 2).

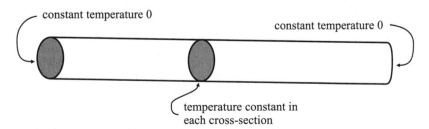

constant temperature 0

constant temperature 0

temperature constant in
each cross-section

Figure 2.   Heat distribution in an insulated rod.

Assume that the endpoints are held at temperature 0, and that the temperature of each cross section is constant. The problem is to describe the temperature $u(x, t)$ of the point x in the rod at time t. Fourier perceived the fundamental importance of this problem as follows:

> Primary causes are unknown to us; but are subject to simple and constant laws, which may be discovered by observation, the study of them being the object of natural philosophy.
>
> Heat, like gravity, penetrates every substance of the universe, its rays occupying all parts of space. The object of our work is to set forth the mathematical laws which this element obeys. The theory of heat will hereafter form one of the most important branches of general physics. . . .
>
> I have deduced these laws from prolonged study and attentive comparison of the facts known up to this time; all these facts I have observed afresh in the course of several years with the most exact instruments that have hitherto been used.

Let us now describe the manner in which Fourier solved his problem. First, it is required to write a differential equation which u satisfies. We shall derive such an equation using three physical principles:

(1.0.4) The density of heat energy is proportional to the temperature u, hence the amount of heat energy in any interval $[a, b]$ of the rod is proportional to $\int_a^b u(x, t)\, dx$.

(1.0.5) **[Newton's Law of Cooling]**   The rate at which heat flows from a hot place to a cold one is proportional to the difference in temperature. The infinitesimal version of this statement is that the rate of heat flow across a point x (from left to right) is some negative constant times $\partial_x u(x, t)$.

(1.0.6) **[Conservation of Energy]**   Heat has no sources or sinks.

Now (1.0.6) tells us that the only way that heat can enter or leave any interval portion $[a, b]$ of the rod is through the endpoints. And (1.0.5) tells us exactly how this happens. Using (1.0.4), we may therefore write

$$\frac{d}{dt}\int_a^b u(x,t)\,dx = \eta^2[\partial_x u(b,t) - \partial_x u(a,t)].$$

We may rewrite this equation as

$$\int_a^b \partial_t u(x,t)\,dx = \eta^2 \int_a^b \partial_x^2 u(x,t)\,dx.$$

Differentiating in b, we find that

$$\partial_t u = \eta^2 \partial_x^2 u, \tag{1.0.7}$$

and that is the heat equation.

Suppose for simplicity that the constant of proportionality η^2 equals 1. Fourier guessed that the equation (1.0.7) has a solution of the form $u(x,t) = \alpha(x)\beta(t)$. Substituting this guess into the equation yields

$$\alpha(x)\beta'(t) = \alpha''(x)\beta(t)$$

or

$$\frac{\beta'(t)}{\beta(t)} = \frac{\alpha''(x)}{\alpha(x)}.$$

Since the left side is independent of x and the right side is independent of t, it follows that there is a constant K such that

$$\frac{\beta'(t)}{\beta(t)} = K = \frac{\alpha''(x)}{\alpha(x)}$$

or

$$\beta'(t) = K\beta(t)$$

$$\alpha''(x) = K\alpha(x).$$

We conclude that $\beta(t) = Ce^{Kt}$. The nature of β, and hence of α, thus depends on the sign of K. But physical considerations tell us that the temperature will dissipate as time goes on, so we conclude that $K \le 0$. Therefore $\alpha(x) = \cos\sqrt{-K}x$ and $\alpha(x) = \sin\sqrt{-K}x$ are solutions of the differential equation for α. The initial conditions $u(0, t) = u(\pi, t) = 0$ (since the ends of the rod are held at constant temperature 0) eliminate the first of these solutions and force $K = -j^2$, $j \in \mathbb{Z}$. Thus Fourier found the solutions

$$u_j(x, t) = e^{-j^2 t} \sin jx, \quad j \in \mathbb{N}$$

of the heat equation. By linearity, any finite linear combination

$$\sum_j b_j e^{-j^2 t} \sin jx$$

of these solutions is also a solution. It is plausible to extend this assertion to infinite linear combinations. Using the initial condition $u(x, 0) = f(x)$ again raises the question of whether "any" function $f(x)$ on $[0, \pi]$ can be written as a (infinite) linear combination of the functions $\sin jx$.

Fourier's solution to this last problem (of the sine functions spanning essentially everything) is roughly as follows. Suppose f is a function that is so representable:

$$f(x) = \sum_j b_j \sin jx. \tag{1.0.8}$$

Setting $x = 0$ gives

$$f(0) = 0.$$

Differentiating both sides of (1.0.8) and setting $x = 0$ gives

$$f'(0) = \sum_{j=1}^{\infty} j b_j.$$

Successive differentiation of (1.0.8), and evaluation at 0, gives

$$f^{(k)}(0) = \sum_{j=1}^{\infty} j^k b_j (-1)^{[k/2]}$$

for k odd (by oddness of f, the even derivatives must be 0 at 0). Here [ ] denotes the greatest integer function. Thus Fourier devised a system of infinitely many equations in the infinitely many unknowns $\{b_j\}$. He proceeded to solve this system by truncating it to an $N \times N$ system (the first N equations restricted to the first N unknowns), solved that truncated system, and then let N tend to ∞. Suffice it to say that Fourier's arguments contained many dubious steps (see [FOU] and [LAN]).

The upshot of Fourier's intricate and lengthy calculations was that

$$b_j = \frac{2}{\pi} \int_0^\pi f(x) \sin jx \, dx. \qquad (1.0.9)$$

By modern standards, Fourier's reasoning was specious; for he began by assuming that f possessed an expansion in terms of sine functions. The formula (1.0.9) hinges on that supposition, together with steps in which one compensated division by zero with a later division by ∞. Nonetheless, Fourier's methods give an actual *procedure* for endeavoring to expand any given f in a series of sine functions.

Fourier's abstract arguments constitute the first part of his book. The bulk, and remainder, of the book consists of separate chapters in which the expansions for particular functions are computed.

Of course we now realize that there is a more direct route to Fourier's formula. If we assume in advance that

$$f(x) = \sum_j b_j \sin jx$$

and that the convergence is in L^2, then we may calculate

$$\frac{2}{\pi} \int_0^\pi f(x) \sin kx \, dx = \frac{2}{\pi} \int_0^\pi \left(\sum_j b_j \sin jx \right) \sin kx \, dx$$

$$= \frac{2}{\pi} \sum_j b_j \int_0^\pi \sin jx \sin kx \, dx$$

$$= b_k.$$

[We use here the fact that $\int \sin jx \sin kx \, dx = 0$ if $j \neq k$, i.e., the fact that $\{\sin jx\}$ are orthogonal in $L^2[0, \pi]$.]

This type of argument was already used by Euler in the late eighteenth century. Of course the notion of a Hilbert space, or of L^2, or of orthogonality of functions, was unknown in Fourier's time. Fourier's discovery, while not rigorous, must be considered an important conceptual breakthrough.

Classical studies of Fourier series were devoted to expanding a function on either $[0, 2\pi)$ or $[0, \pi)$ in a series of the form

$$\sum_{j=0}^{\infty} a_j \cos jx$$

or

$$\sum_{j=1}^{\infty} b_j \sin jx$$

or as a combination of these

$$\sum_{j=0}^{\infty} a_j \cos jx + \sum_{j=1}^{\infty} b_j \sin jx.$$

The modern theory tends to use the more elegant notation of complex exponentials. Since

$$\cos jx = \frac{e^{ijx} + e^{-ijx}}{2} \qquad \text{and} \qquad \sin jx = \frac{e^{ijx} - e^{-ijx}}{2i},$$

we may seek to instead expand f in a series of the form

$$\sum_{j=-\infty}^{\infty} c_j e^{ijx}.$$

In this book we shall confine ourselves to the exponential notation.

Derivation of the Wave Equation

We imagine a perfectly flexible elastic string with negligible weight. Our analysis will ignore damping effects, such as air resistance. We

Figure 3.   The string in its relaxed position.

assume that, in its relaxed position, the string is as in Figure 3. The
string is plucked in the vertical direction, and is thus set in motion in a
vertical plane.

We focus attention on an "element" Δx of the string (Figure 4)
that lies between x and $x + \Delta x$. We adopt the usual physical conceit of
assuming that the displacement (motion) of this string element is *small*,
so that there is only a slight error in supposing that the motion of each
point of the string element is strictly vertical. We let the tension of the
string, at the point x at time t, be denoted by $T(x, t)$. Note that T acts

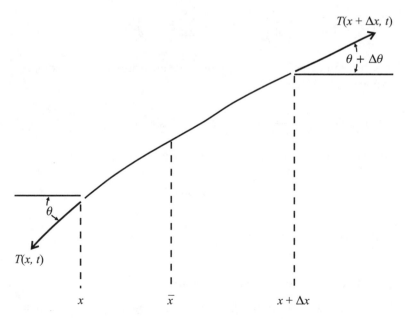

Figure 4.   An element of the string.

only in the tangential direction (i.e., along the string). We denote the mass density of the string by ρ.

Since *there is no horizontal component of acceleration*, we see that

$$T(x + \Delta x, t) \cdot \cos(\theta + \Delta\theta) - T(x, t) \cdot \cos(\theta) = 0. \qquad (1.0.10)$$

[Refer to Figure 5: The expression $T(\star) \cdot \cos(\star)$ denotes $H(\star)$, the horizontal component of the tension.] Thus equation (1.0.10) says that H is independent of x.

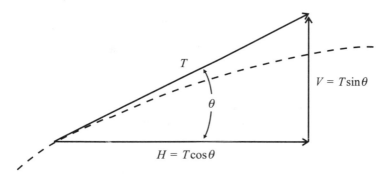

Figure 5.   Tension on the string.

Now we look at the vertical component of force (acceleration):

$$T(x+\Delta x, t) \cdot \sin(\theta+\Delta\theta) - T(x, t) \cdot \sin(\theta) = \rho \cdot \Delta x \cdot u_{tt}(\overline{x}, t). \quad (1.0.11)$$

Here $\overline{x}$ is the mass center of the string element and we are applying Newton's second law—that the external force is the mass of the string element times the acceleration of its center of mass. We use subscripts to denote derivatives. We denote the vertical component of $T(\star)$ by $V(\star)$. Thus equation (1.0.11) can be written as

$$\frac{V(x + \Delta x, t) - V(x, t)}{\Delta x} = \rho \cdot u_{tt}(x, t).$$

Letting $\Delta x \to 0$ yields

$$V_x(x, t) = \rho \cdot u_{tt}(x, t). \tag{1.0.12}$$

We would like to express equation (1.0.12) entirely in terms of u, so we notice that

$$V(x, t) = H(t) \tan \theta = H(t) \cdot u_x(x, t).$$

[We have used the fact that the derivative in x is the slope of the tangent line, which is $\tan \theta$.] Substituting this expression for V into (1.0.12) yields

$$(H u_x)_x = \rho \cdot u_{tt}.$$

But H is independent of x, so this last line simplifies to

$$H \cdot u_{xx} = \rho \cdot u_{tt}.$$

For small displacements of the string, θ is nearly zero, so $H = T \cos \theta$ is nearly T. Thus we finally write our equation as

$$\frac{T}{\rho} u_{xx} = u_{tt}.$$

It is traditional to denote the constant on the left by a^2. We finally arrive at the *wave equation*

$$a^2 u_{xx} = u_{tt}.$$

1.1   The Rudiments of Fourier Series

Euclidean spaces are equipped with special groups that act on them in a natural geometric fashion. For instance $\mathbb{R}^N$, $N \geq 2$ is acted on by translations, rotations, and dilations. It is natural to study functions on such a space that transform nicely under the group action. In fact these considerations lead to the definition of the Fourier transform, which we study in Chapter 2.

In the present chapter we study the circle group, which is formally defined as $\mathbb{T} = \mathbb{R}/2\pi\mathbb{Z}$ (see also Section 0.3). What this quotient means is that we consider the equivalence classes of real numbers that are induced by the equivalence relation $x \sim y$ if $x - y$ is divisible by 2π. A useful model for $\mathbb{T}$ is the interval $[0, 2\pi)$ with addition performed modulo 2π. When we think of a function f on $\mathbb{T}$, we naturally identify it with its 2π-periodic extension to the real line. Therefore, when we integrate f, we are free to integrate it from any real number b to $b + 2\pi$; the value will be independent of the choice of b.

It is also sometimes useful to identify the circle group with the unit circle S in the complex plane. We do so with the mapping

$$[0, 2\pi) \ni x \longmapsto e^{ix} \in S. \tag{1.1.1}$$

Clearly the circle group has only itself acting on it naturally as a group. That is to say, if g is a fixed element of the circle group, then it induces the map

$$\tau_g : [0, 2\pi) \ni x \longmapsto x + g \in [0, 2\pi),$$

where again we are performing addition modulo 2π. Let us consider functions which transform naturally under this group action.

Of course if we were to require that a function f on the circle group literally *commute* with translation, then f would be constant. It turns out to be more natural to require that there exist a function ϕ, with $|\phi(x)| = 1$ for all x, such that

$$f(y + x) = \phi(x) \cdot f(y). \tag{1.1.2}$$

Thus the *size* of f is preserved under the group action. Taking $y = 0$ in (1.1.2) yields

$$f(x) = \phi(x) \cdot f(0),$$

so we see right away that f is completely determined by its value at 0 and by the factor function ϕ. In addition, we compute that

$$\phi(x + y) \cdot f(0) = f(x + y) = \phi(x) \cdot f(0 + y) = \phi(x) \cdot \phi(y) \cdot f(0).$$

Thus, as long as f is not the identically zero function, we see that

$$\phi(x + y) = \phi(x) \cdot \phi(y). \tag{1.1.3}$$

Our conclusion is this: any function ϕ which satisfies the transformation law (1.1.2) for some function f must have property (1.1.3). If, in addition, $|\phi| = 1$, then (1.1.3) says that ϕ must be a group homomorphism from the circle group into the unit circle S in the complex plane. [The calculations that we have just performed are taken from [FOL1]. The book [FOL2] is also an excellent source for information about Fourier analysis on locally compact abelian groups.]

In fact the preceding computations are valid on any compact abelian group—where a topological group is a group equipped with a topology that makes the group operations continuous. So let us step back for a minute and describe the elements of Fourier analysis on such a group.

When studying the Fourier analysis of a locally compact abelian group G, one begins by classifying all the continuous homomorphisms $\phi : G \to \mathbb{C}^*$, where $\mathbb{C}^*$ is the group $\mathbb{C} \setminus \{0\}$ under multiplication. These mappings are called the *group characters*; the characters themselves form a group, and they are the building blocks of commutative Fourier analysis. The functions ϕ that we discovered in line (1.1.3) are the characters of $\mathbb{T}$.

If our group G is compact, then it is easy to see that any character ϕ must have image lying in the unit circle. For the image of ϕ must be compact. If $\lambda = \phi(g)$ is in the image of ϕ and has modulus greater than 1, then $\lambda^k = \phi(g^k)$ will tend to ∞ as $\mathbb{N} \ni k \to +\infty$. That contradicts the compactness of the image. A similar contradiction results if $|\lambda| < 1$. It follows that the image of ϕ must lie in the unit circle.

We shall now find all the characters of the group $\mathbb{T}$ (we thank R. Burckel for this argument; see also [FOL1, p. 239]). In this calculation, we freely identify the interval $[0, 2\pi)$ with $\mathbb{T}$ by way of the map $t \mapsto e^{it}$ (as already suggested in line (1.1.1)). Now suppose that ϕ is a character. So ϕ is a continuous function on $[0, 2\pi)$ and $\phi(s + t) = \phi(s) \cdot \phi(t)$. In particular, $\phi(0) = 1$.

We may select a number $0 < a < 2\pi$ such that $\int_0^a \phi(t)\,dt \neq 0$. Let b denote the value of this integral. Then

$$
\begin{aligned}
b \cdot \phi(x) &= \phi(x) \cdot \int_0^a \phi(t)\,dt \\[2mm]
&= \int_0^a \phi(x) \cdot \phi(t)\,dt \\[2mm]
&= \int_0^a \phi(x + t)\,dt \\[2mm]
&\overset{(s=x+t)}{=} \int_x^{x+a} \phi(s)\,ds.
\end{aligned}
$$

It follows that

$$
\phi(x) = \frac{1}{b} \int_x^{x+a} \phi(s)\,ds;
$$

hence ϕ is continuously differentiable. Therefore we may calculate:

$$
\phi'(x) = \lim_{h \to 0} \frac{\phi(x + h) - \phi(x)}{h} = \lim_{h \to 0} \frac{\phi(x) \cdot \phi(h) - \phi(x)}{h}
$$

$$
= \phi(x) \cdot \lim_{h \to 0} \frac{\phi(h) - \phi(0)}{h} = \phi(x) \cdot \phi'(0).
$$

We write this equation as

$$
\phi'(x) = K \cdot \phi(x)
$$

with $K = \phi'(0)$.

We solve (or "integrate") this equation by multiplying through by e^{-Kx}:

$$
0 = \phi'(x)e^{-Kx} - K\phi(x)e^{-Kx}.
$$

In other words

$$
0 = \frac{d}{dx}\left[\phi(x)e^{-Kx} \right];
$$

hence

$$
\phi(x)e^{-Kx} = c
$$

or

$$\phi(x) = c \cdot e^{Kx}.$$

Because $\phi(0) = 1$ we know that $c = 1$. Because ϕ is bounded by 1 we know that K must be pure imaginary. Therefore, for some $r \in \mathbb{R}$,

$$\phi(x) = e^{irx}, \qquad \text{all } x \in \mathbb{R}.$$

Because ϕ is periodic, it must be that $r = k \in \mathbb{Z}$. We conclude that the group of characters of $\mathbb{T}$ is isomorphic to the additive group of integers.

Now suppose that f is a function of the form

$$f(t) = \sum_{j=-N}^{N} a_j e^{ijt}.$$

We call such a function a *trigonometric polynomial*. Trigonometric polynomials are dense in $L^p(\mathbb{T})$, $1 \leq p < \infty$ (see Appendix I). In that sense they are "typical" functions. Notice that, if $-N \leq k \leq N$, then

$$\frac{1}{2\pi} \int_0^{2\pi} f(t) e^{-ikt} \, dt = \sum_{j=-N}^{N} a_j \frac{1}{2\pi} \int_0^{2\pi} e^{ijt} e^{-ikt} \, dt = a_k.$$

This calculation shows that we may recover the kth *Fourier coefficient* of f by integrating f against the conjugate of the corresponding character. This consideration leads us to the following definition:

Definition 1.1.4 Let f be an integrable function on $\mathbb{T}$. For $j \in \mathbb{Z}$, we define

$$\widehat{f}(j) = a_j \equiv \frac{1}{2\pi} \int_0^{2\pi} f(t) e^{-ijt} \, dt.$$

We call $\widehat{f}(j) = a_j$ the jth *Fourier coefficient* of f.

In the subject of Fourier series, it is convenient to build a factor of $1/2\pi$ into our integrals. We have already seen this feature in the definition of the Fourier coefficients. But we will also let

$$\|f\|_{L^p(\mathbb{T})} \equiv \left[\frac{1}{2\pi} \int_0^{2\pi} |f(e^{it})|^p \, dt \right]^{1/p}, \quad 1 \le p < \infty.$$

The fundamental issue of Fourier analysis is this: We introduce the formal expression

$$Sf \sim \sum_{j=-\infty}^{\infty} \widehat{f}(j) e^{ijt}. \tag{1.1.5}$$

We call the expression *formal*, because we do not know whether the series converges; if it does converge, we do not know whether it converges[1] to the function f.

It will turn out that Fourier series are much more cooperative than Taylor series and other types of series in analysis. For very broad classes of functions, the Fourier series is at least summable (a concept to be defined later) to f. Thus it is indeed possible to realize Daniel Bernoulli's dream: to represent an arbitrary input for the wave equation as a superposition of sine and cosine functions (remember that $e^{ijt} = \cos jt + i \sin jt$.)

We conclude this section with three basic results about the size of Fourier coefficients.

Proposition 1.1.6 *Let f be an integrable function on $\mathbb{T}$. Then, for each integer j,*

$$|\widehat{f}(j)| \le \frac{1}{2\pi} \int |f(t)| \, dt.$$

In other words,

$$|\widehat{f}(j)| \le \|f\|_{L^1}.$$

---

[1] Recall here the theory of Taylor series from calculus: The Taylor series for a typical C^∞ function g generally does not converge, and when it does converge it does not typically converge to the function g.

Proof. We observe that

$$|\widehat{f}(j)| = \left| \frac{1}{2\pi} \int_0^{2\pi} f(t)e^{-ijt}\, dt \right| \le \frac{1}{2\pi} \int_0^{2\pi} |f(t)|\, dt,$$

as was to be proved.   □

Proposition 1.1.7 (Riemann-Lebesgue) *Let f be an integrable function on $\mathbb{T}$. Then*

$$\lim_{j \to \pm\infty} |\widehat{f}(j)| = 0.$$

Proof. First consider the case when f is a trigonometric polynomial. Say that

$$f(t) = \sum_{j=-M}^{M} a_j e^{ijt}.$$

Then $\widehat{f}(j) = 0$ as soon as $|j| \ge M$. That proves the result for trigonometric polynomials.

Now let f be any integrable function. Let $\epsilon > 0$. Choose a trigonometric polynomial p such that $\|f - p\|_{L^1} < \epsilon$ (see Appendix I). Let N be the degree of the trigonometric polynomial p and let $|j| > N$. Then

$$\begin{aligned}|\widehat{f}(j)| &\le |[f - p]\widehat{}(j)| + |\widehat{p}(j)| \\ &\le \|f - p\|_{L^1} + 0 \\ &< \epsilon.\end{aligned}$$

That proves the result.   □

Proposition 1.1.8 *Let f be a k times continuously differentiable 2π-periodic function. Then the Fourier coefficients of f satisfy*

$$|\widehat{f}(j)| \le C_k \cdot (1 + |j|)^{-k}.$$

Sketch of Proof. We know that, for $j \neq 0$,

$$\widehat{f}(j) \quad = \quad \frac{1}{2\pi} \int_0^{2\pi} f(t)e^{-ijt}\, dt$$

$$\overset{\text{(parts)}}{=} \quad \frac{1}{2\pi} \frac{1}{ij} \int_0^{2\pi} f'(t)e^{-ijt}\, dt$$

$$= \cdots =$$

$$\overset{\text{(parts)}}{=} \quad \frac{1}{2\pi} \frac{1}{(ij)^k} \int_0^{2\pi} f^{(k)}(t)e^{-ijt}\, dt.$$

Notice that the boundary terms vanish by the periodicity of f. Thus

$$|\widehat{f}(j)| \leq C \cdot \frac{|\widehat{f^{(k)}}(j)|}{|j|^k} \leq \frac{C'}{|j|^k} \leq \frac{2^k C'}{(1+|j|)^k},$$

where

$$C' = C \cdot \frac{1}{2\pi} \int |f^{(k)}(x)|\, dx = C \cdot \|f^{(k)}\|_{L^1}.$$

This is the desired result.   □

This last result has a sort of converse: If the Fourier coefficients of a function decay rapidly, then the function is smooth. Indeed, the more rapid the decay of the Fourier coefficients, the smoother the function. This circle of ideas continues to be an active area of research, and currently is being studied in the context of wavelet theory (see Chapter 7).

1.2   Summability of Fourier Series

Hot debate over the summing of Fourier series was brought to a halt by P. Dirichlet in 1828; for the first time in the context of Fourier series, he defined what was meant for a series to *converge*. Given our current perspective, it is clear that the first step in such a program is to define the notion of a partial sum:

Definition 1.2.1 Let f be an integrable function on $\mathbb{T}$ and let the formal Fourier series of f be as in (1.1.4). We define the Nth *partial sum* of f to be the expression

$$S_N f(x) = \sum_{j=-N}^{N} \widehat{f}(j) e^{ijx}.$$

We say that the Fourier series *converges* to f at the point x if

$$S_N f(x) \to f(x) \qquad \text{as } N \to \infty$$

in the sense of convergence of ordinary sequences of complex numbers.

It is most expedient to begin our study of summation of Fourier series by finding an integral formula for $S_N f$. Thus we write

$$S_N f(x) = \sum_{j=-N}^{N} \widehat{f}(j) e^{ijx}$$

$$= \sum_{j=-N}^{N} \frac{1}{2\pi} \int_0^{2\pi} f(t) e^{-ijt} \, dt \, e^{ijx}$$

$$= \frac{1}{2\pi} \int_0^{2\pi} \left[\sum_{j=-N}^{N} e^{ij(x-t)} \right] f(t) \, dt. \qquad (1.2.2)$$

We need to calculate the sum in brackets; for that will be a universal object associated to the summation process S_N, and unrelated to the particular function f that we are considering.

Now

$$\sum_{j=-N}^{N} e^{ijs} = e^{-iNs} \sum_{j=0}^{2N} e^{ijs}$$

$$= e^{-iNs} \sum_{j=0}^{2N} [e^{is}]^j. \qquad (1.2.3)$$

The sum on the right is a geometric sum, and we may instantly write a formula for it (as long as $s \neq 0 \bmod 2\pi$):

$$\sum_0^{2N} [e^{is}]^j = \frac{e^{i(2N+1)s} - 1}{e^{is} - 1}.$$

Substituting this expression into (1.2.3) yields

$$\sum_{j=-N}^N e^{ijs} = e^{-iNs} \frac{e^{i(2N+1)s} - 1}{e^{is} - 1}$$

$$= \frac{e^{i(N+1)s} - e^{-iNs}}{e^{is} - 1}$$

$$= \frac{e^{i(N+1)s} - e^{-iNs}}{e^{is} - 1} \cdot \frac{e^{-is/2}}{e^{-is/2}}$$

$$= \frac{e^{i(N+1/2)s} - e^{-i(N+1/2)s}}{e^{is/2} - e^{-is/2}}$$

$$= \frac{\sin \left[N + \frac{1}{2} \right] s}{\sin \frac{1}{2} s}.$$

We see that we have derived a closed formula (no summation signs) for the relevant sum. In other words, using (1.2.2), we now know that

$$S_N f(x) = \frac{1}{2\pi} \int_0^{2\pi} \frac{\sin \left[N + \frac{1}{2} \right] (x - t)}{\sin \frac{x-t}{2}} f(t) \, dt.$$

The expression

$$D_N(s) = \frac{\sin \left[N + \frac{1}{2} \right] s}{\sin \frac{s}{2}}$$

is called the *Dirichlet kernel*. It is the fundamental object in any study of the summation of Fourier series. In summary, our formula is

$$S_N f(x) = \frac{1}{2\pi} \int_0^{2\pi} D_N(x - t) f(t) \, dt$$

or (after a change of variable—where we exploit the fact that our functions are periodic to retain the same limits of integration)

$$S_N f(x) = \frac{1}{2\pi} \int_0^{2\pi} D_N(t) f(x - t) \, dt.$$

For now, we notice that

$$\frac{1}{2\pi} \int_0^{2\pi} D_N(t) \, dt = \frac{1}{2\pi} \int_0^{2\pi} \sum_{j=-N}^{N} e^{ijt} \, dt$$

$$= \sum_{j=-N}^{N} \frac{1}{2\pi} \int_0^{2\pi} e^{ijt} \, dt$$

$$= \frac{1}{2\pi} \int_0^{2\pi} e^{i0t} \, dt$$

$$= 1.$$

Departing from many popular studies, we will begin our work with the following not-terribly-well-known theorem:

Theorem 1.2.4 *Let f be an integrable function on $\mathbb{T}$ and suppose that f is differentiable at x. Then $S_N f(x) \to f(x)$.*

An immediate corollary of the theorem is that the Fourier series of a differentiable function converges to that function at *every* point. Contrast this result with the situation for Taylor series!

Proof of the Theorem. We examine the expression $S_N f(x) - f(x)$:

$$|S_N f(x) - f(x)|$$

$$= \left| \frac{1}{2\pi} \int_0^{2\pi} D_N(t) f(x - t) \, dt - f(x) \right|$$

$$= \left| \frac{1}{2\pi} \int_0^{2\pi} D_N(t) f(x - t) \, dt - \frac{1}{2\pi} \int_0^{2\pi} D_N(t) f(x) \, dt \right|.$$

Notice that something very important has transpired in the last step: We used the fact that $\frac{1}{2\pi} \int_0^{2\pi} D_N(t)\,dt = 1$ to rewrite the simple expression $f(x)$ (which is constant with respect to the variable t) in an interesting fashion; this step will allow us to combine the two expressions inside the absolute value signs.

Thus we have

$$|S_N f(x) - f(x)| = \left| \frac{1}{2\pi} \int_0^{2\pi} D_N(t)[f(x-t) - f(x)]\,dt \right|.$$

We may translate f so that $x = 0$, and (by periodicity) we may perform the integration from $-\pi$ to π (instead of from 0 to 2π). Thus our integral is

$$P_N \equiv \left| \frac{1}{2\pi} \int_{-\pi}^{\pi} D_N(t)[f(t) - f(0)]\,dt \right|.$$

Note that another change of variable has allowed us to replace $-t$ by t. Now fix $\epsilon > 0$ and write

$$P_N \le \left\{ \left| \frac{1}{2\pi} \int_{-\pi}^{-\epsilon} D_N(t)[f(t) - f(0)]\,dt \right| \right.$$

$$+ \left. \left| \frac{1}{2\pi} \int_{\epsilon}^{\pi} D_N(t)[f(t) - f(0)]\,dt \right| \right\}$$

$$+ \left| \frac{1}{2\pi} \int_{-\epsilon}^{\epsilon} D_N(t)[f(t) - f(0)]\,dt \right|$$

$$\equiv I + II.$$

We may note that

$$\sin(N + 1/2)t = \sin Nt \cos t/2 + \cos Nt \sin t/2$$

and thus rewrite I as

$$\left| \frac{1}{2\pi} \int_{-\pi}^{-\epsilon} \sin Nt \left[\cos \frac{1}{2}t \cdot \frac{f(t) - f(0)}{\sin \frac{t}{2}} \right] dt \right.$$

$$+ \frac{1}{2\pi} \int_{-\pi}^{-\epsilon} \cos Nt \left[\sin \frac{1}{2}t \cdot \frac{f(t) - f(0)}{\sin \frac{t}{2}} \right] dt \Bigg|$$

$$+ \left| \frac{1}{2\pi} \int_{\epsilon}^{\pi} \sin Nt \left[\cos \frac{1}{2}t \cdot \frac{f(t) - f(0)}{\sin \frac{t}{2}} \right] dt \right.$$

$$+ \frac{1}{2\pi} \int_{\epsilon}^{\pi} \cos Nt \left[\sin \frac{1}{2}t \cdot \frac{f(t) - f(0)}{\sin \frac{t}{2}} \right] dt \Bigg|.$$

These four expressions are all analyzed in the same way, so let us look at the first of them. The expression

$$\left[\cos \frac{1}{2}t \cdot \frac{f(t) - f(0)}{\sin \frac{t}{2}} \right] \chi_{[-\pi, -\epsilon]}(t)$$

is an integrable function (because t is bounded from zero on its support). Call it $g(t)$. Then our first integral may be written as

$$\left| \frac{1}{2\pi} \frac{1}{2i} \int_{-\pi}^{\pi} e^{iNt} g(t)\, dt - \frac{1}{2\pi} \frac{1}{2i} \int_{-\pi}^{\pi} e^{-iNt} g(t)\, dt \right|.$$

Each of these last two expressions is ($1/2i$ times) the $\pm N^{\text{th}}$ Fourier coefficient of the integrable function g. The Riemann-Lebesgue lemma tells us that, as $N \to \infty$, they tend to zero. That takes care of I.

The analysis of II is similar, but slightly more delicate. First observe that

$$f(t) - f(0) = \mathcal{O}(t).$$

[Here $\mathcal{O}(t)$ is Landau's notation for an expression that is not greater than $C \cdot |t|$—see Appendix IX.] More precisely, the differentiability of f at 0 means that $[f(t) - f(0)]/t \to f'(0)$, hence $|f(t) - f(0)| \leq C \cdot |t|$ for t small.

Thus

$$II = \left| \frac{1}{2\pi} \int_{-\epsilon}^{\epsilon} \frac{\sin \left[N + \frac{1}{2} \right] t}{\sin \frac{t}{2}} \cdot \mathcal{O}(t)\, dt \right|.$$

Regrouping terms, as we did in our estimate of I, we see that

$$II = \left| \frac{1}{2\pi} \int_{-\epsilon}^{\epsilon} \sin Nt \left[\cos \frac{t}{2} \cdot \frac{\mathcal{O}(t)}{\sin \frac{t}{2}} \right] dt \right|$$

$$+ \left| \frac{1}{2\pi} \int_{-\epsilon}^{\epsilon} \cos Nt \left[\sin \frac{t}{2} \cdot \frac{\mathcal{O}(t)}{\sin \frac{t}{2}} \right] dt \right|.$$

The expressions in brackets are integrable functions (in the first instance, because $\mathcal{O}(t)$ cancels the singularity that would be induced by $\sin[t/2]$), and (as before) integration against $\cos Nt$ or $\sin Nt$ amounts to calculating a $\pm N^{\text{th}}$ Fourier coefficient. As $N \to \infty$, these tend to zero by the Riemann-Lebesgue lemma.

To summarize, our expression P_N tends to 0 as $N \to \infty$. That is what we wished to prove. $\qquad\qquad\qquad\qquad\qquad\qquad\qquad\qquad$ $\square$

1.3 A Quick Introduction to Summability Methods

For many practical applications, the result presented in Theorem 1.2 is sufficient. Many "calculus-style" functions that we encounter in practice are differentiable except at perhaps finitely many points (we call these *piecewise differentiable functions*). The theorem guarantees that the Fourier series of such a function will converge back to the function except perhaps at those finitely many singular points. A standard—and very useful—theorem of Fejér provides the further information that, if f is differentiable except at finitely many jump discontinuities, then the Fourier series of f converges to $\frac{1}{2}[f(x+) + f(x-)]$ at each point x. Another refinement is this: the conclusion of Theorem 1.2.4 holds if the function f merely satisfies a suitable Lipschitz condition at the point x.

However, for other purposes, one wishes to treat an entire Banach space of functions—for instance L^2 or L^p. Pointwise convergence for the Fourier series of a function in one of these spaces is the famous

Carleson-Hunt theorem [CAR], [HUN], one of the deepest results in all of modern analysis. We certainly cannot treat it here. "Summability" is much easier, and in practice is just as useful. We can indeed explain the basic ideas of summability in this brief treatment.

In order to obtain a unified approach to various summability methods, we shall introduce some ancillary machinery. This will involve some calculation, and some functional analysis. Our approach is inspired by [KAT]. Let us begin at the beginning: We shall give two concrete examples of summability methods, and explain what they are.

As we noted previously, one establishes ordinary convergence of a Fourier series by examining the sequence of partial summation operators $\{S_N\}$. Figure 1 exhibits the "profile" of the operator S_N. In technical language, this figure exhibits the *Fourier multiplier* associated to the operator S_N. More generally, let f be an integrable function and $\sum_{-\infty}^{\infty} \widehat{f}(j)e^{ijx}$ its (formal) Fourier series. If $\Lambda = \{\lambda_j\}_{j=-\infty}^{\infty}$ is a sequence of complex numbers, then Λ acts as a Fourier multiplier according to the rule

$$\mathcal{M}_\Lambda : f \longmapsto \sum_j \lambda_j \widehat{f}(j)e^{ijx}.$$

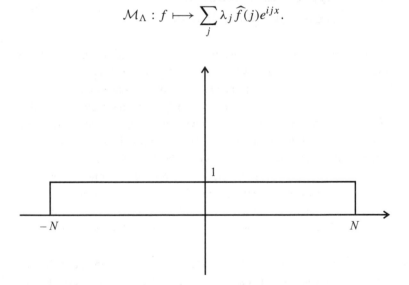

Figure 1. The "profile" of S_N.

In this language, the multiplier

$$\lambda_j = \begin{cases} 1 & \text{if} \quad |j| \le N; \\ 0 & \text{if} \quad |j| > N \end{cases}$$

corresponds to the partial-summation operator S_N. The picture of the multiplier, shown in Figure 1, enables us to see that the multiplier exhibits a precipitous "drop" at $\pm N$: The multiplier has a sudden change of value from 1 to 0.

The spirit of summability methods, as we now understand them, is to average the partial-summation operators in such a way as to mollify the sharp drop of the multipliers corresponding to the operators S_N. Fejér's method for achieving this effect—now known as a special case of the Cesàro summation method—is as follows. For f an integrable function, we define

$$\sigma_N f(x) = \frac{1}{N+1} \sum_{j=0}^{N} S_j f(x).$$

Notice that we are simply averaging the first $N + 1$ partial-summation operators for f. Just as we calculated a closed formula for $S_N f$, let us now calculate a closed formula for $\sigma_N f$.

If we let K_N denote the kernel of σ_N, then we find that

$$K_N(x) = \frac{1}{N+1} \sum_{j=0}^{N} D_j f(x)$$

$$= \frac{1}{N+1} \sum_{j=0}^{N} \frac{\sin\left[j + \frac{1}{2}\right]x}{\sin\frac{x}{2}}$$

$$= \frac{1}{N+1} \sum_{j=0}^{N} \frac{\cos jx - \cos(j+1)x}{2\sin^2\frac{x}{2}}$$

(since $\sin a \sin b = \frac{1}{2}[\cos(a - b) - \cos(a + b)]$). Of course the sum collapses and we find that

$$K_N(x) = \frac{1}{N+1} \frac{1 - \cos(N+1)x}{2\sin^2 \frac{x}{2}}$$

$$= \frac{1}{N+1} \frac{1 - [\cos^2(\frac{(N+1)x}{2}) - \sin^2(\frac{(N+1)x}{2})]}{2\sin^2 \frac{x}{2}}$$

$$= \frac{1}{N+1} \frac{2\sin^2(\frac{(N+1)x}{2})}{2\sin^2 \frac{x}{2}}$$

$$= \frac{1}{N+1} \left(\frac{\sin(\frac{(N+1)x}{2})}{\sin \frac{x}{2}} \right)^2 .$$

Notice that the Fourier multiplier associated to Fejér's summation method is

$$\lambda_j = \begin{cases} \frac{N+1-|j|}{N+1} & \text{if} \quad |j| \le N \\ 0 & \text{if} \quad |j| > N. \end{cases}$$

We can see that this multiplier effects the transition from 1 to 0 gradually, over the range $|j| \le N$. Contrast this with the multiplier associated to ordinary partial summation.

On the surface, it may not be apparent why K_N is a more useful and accessible kernel than D_N, but we shall attend to those details shortly. Before we do so, let us look at another summability method for Fourier series. This method is due to Poisson, and is now understood to be a special instance of the summability method of Abel.

For f an integrable function and $0 < r < 1$ we set

$$P_r f(x) = \sum_{j=-\infty}^{\infty} r^{|j|} \widehat{f}(j) e^{ijx}.$$

Notice now that the Fourier multiplier is $\Lambda = \{r^{|j|}\}$. Again the Fourier multiplier exhibits a smooth transition from 1 to 0—in contrast with the multiplier for the partial-summation operator S_N.

Let us calculate the kernel associated to the Poisson summation method. It will be given by

$$P_r(e^{it}) = \sum_{j=-\infty}^{\infty} r^{|j|} e^{ijt}$$

$$= \sum_{j=0}^{\infty} [re^{it}]^j + \sum_{j=0}^{\infty} [re^{-it}]^j - 1$$

$$= \frac{1}{1 - re^{it}} + \frac{1}{1 - re^{-it}} - 1$$

$$= \frac{2 - 2r\cos t}{|1 - re^{it}|^2} - 1$$

$$= \frac{1 - r^2}{1 - 2r\cos t + r^2}.$$

We see that we have rediscovered the familiar Poisson kernel of complex function theory and harmonic analysis. Observe that, for fixed r (or, more generally, for r in a compact subinterval of $[0, 1)$), the series converges uniformly to the Poisson kernel.

It will be by examining the kernels associated to these new multipliers that we will see how they are more effective than partial summation. The results that we will be presenting can be seen to be special cases of (the philosophy of) the Marcinkiewicz multiplier theorem—see [STE1].

Now let us summarize what we have learned, or are about to learn. Let f be an integrable function on the group $\mathbb{T}$.

A. Ordinary partial summation of the Fourier series for f, which is the operation

$$S_N f(x) = \sum_{j=-N}^{N} \widehat{f}(j) e^{ijx},$$

is given by the integral formulas

$$S_N f(x) = \frac{1}{2\pi} \int_0^{2\pi} f(t) D_N(x - t)\, dt$$

$$= \frac{1}{2\pi} \int_0^{2\pi} f(x - t) D_N(t)\, dt,$$

where

$$D_N(t) = \frac{\sin\left[N + \frac{1}{2}\right]t}{\sin\frac{1}{2}t}.$$

B. Fejér summation of the Fourier series for f, a special case of Cesàro summation, is given by

$$\sigma_N f(x) = \frac{1}{N+1} \sum_{j=0}^{N} S_j f(x).$$

It is also given by the integral formulas

$$\sigma_N f(x) = \frac{1}{2\pi} \int_0^{2\pi} f(t) K_N(x - t)\, dt$$

$$= \frac{1}{2\pi} \int_0^{2\pi} f(x - t) K_N(t)\, dt,$$

where

$$K_N(t) = \frac{1}{N+1} \left(\frac{\sin \frac{(N+1)}{2} t}{\sin \frac{t}{2}} \right)^2.$$

C. Poisson summation of the Fourier series for f, a special case of Abel summation, is given by

$$P_r f(x) = \frac{1}{2\pi} \int_0^{2\pi} f(t) P_r(x - t)\, dt = \frac{1}{2\pi} \int_0^{2\pi} f(x - t) P_r(t)\, dt,$$

where

$$P_r(t) = \frac{1 - r^2}{1 - 2r \cos t + r^2}, \qquad 0 \le r < 1.$$

Figure 2 shows the graphs of the Dirichlet, Fejer, and Poisson kernels.

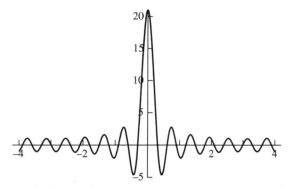

The Dirichlet kernel with $N = 10$

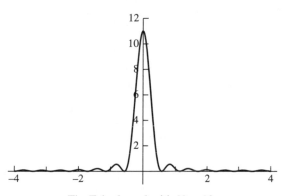

The Fejer kernel with $N = 10$

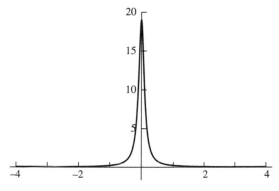

The Poisson kernel with $r = .9$

Figure 2. Graphs of the Dirichlet, Fejer, and Poisson kernels.

In the next section we shall isolate properties of the summability kernels P_r and K_N that make their study direct and efficient.

1.4 Key Properties of Summability Kernels

It is an interesting diversion to calculate

$$\frac{1}{2\pi} \int_0^{2\pi} |D_N(t)|\, dt = \|D_N\|_{L^1}.$$

As an exercise, endeavor to do so by either (i) using `Mathematica` or `Maple` or (ii) breaking up the interval $[0, \pi]$ into subintervals on which $\sin[N + \frac{1}{2}]t$ is essentially constant. By either method, you will find that

$$\|D_N\|_{L^1} \approx c \cdot \log N$$

(here the notation '$\approx$' means "is of the size"). This nonuniform integrability of the Dirichlet kernel is, for the moment, a roadblock to our understanding of the partial-summation process. The theory of singular integral operators will give us a method for handling integral kernels like D_N. We shall say more about these in Sections 1.6 and 1.7 as well as in Chapters 5 and 6.

Meanwhile, let us isolate the properties of summability kernels that distinguish them from D_N and make them most useful. Call the kernels $\{k_N\}_{N \in \mathbb{Z}^+}$. We will consider asymptotic properties of these kernels as $N \to +\infty$. [Note that, in most of the examples we shall present, the indexing space for the summability kernels will be $\mathbb{Z}^+$, the nonnegative integers—just as we have stated. But in some examples, such as the Poisson kernel, it will be more convenient to let the parameter be $r \in [0, 1)$. In this last case we shall consider asymptotic properties of the kernels as $r \to 1^-$. We urge the reader to be flexible about this notation.] There are three properties that are desirable for a family of summability kernels:

(1.4.1) $\dfrac{1}{2\pi} \displaystyle\int_0^{2\pi} k_N(x)\, dx = 1 \qquad \forall N;$

(1.4.2) $\dfrac{1}{2\pi} \displaystyle\int_0^{2\pi} |k_N(x)|\,dx \le C \qquad \forall N, \quad$ some finite $C > 0$;

(1.4.3) If $\epsilon > 0$, then $\displaystyle\lim_{N\to\infty} k_N(x) = 0$, uniformly for $\pi \ge |x| \ge \epsilon$.

Let us call any family of summability kernels *standard* if it possesses these three properties. [See [KAT] for the genesis of some of these ideas.]

It is worth noting that condition (1.4.1) plus positivity of the kernel automatically give condition (1.4.2). Positive kernels will prove to be "friendly" in other respects as well. Both the Fejér and the Poisson kernels are positive.

Now we will check that the family of Fejér kernels and the family of Poisson kernels both possess these three properties.

The Fejér kernels:
Notice that, since $K_N \ge 0$,

$$\frac{1}{2\pi} \int_0^{2\pi} |K_N(t)|\,dt = \frac{1}{2\pi} \int_0^{2\pi} K_N(t)\,dt$$

$$= \frac{1}{N+1} \sum_{j=0}^{N} \frac{1}{2\pi} \int_0^{2\pi} D_j(t)\,dt$$

$$= \frac{1}{N+1} \cdot (N+1) = 1.$$

This takes care of (1.4.1) and (1.4.2). For (1.4.3), notice that $|\sin(t/2)| \ge |\sin(\epsilon/2)| > 0$ when $\pi \ge |t| \ge \epsilon > 0$. Thus, for such t,

$$|K_N(t)| \le \frac{1}{N+1} \cdot \frac{1}{|\sin(\epsilon/2)|} \to 0,$$

uniformly in t as $N \to \infty$. Thus the Fejér kernels form a standard family of summability kernels.

The Poisson kernels:
First we observe that, since $P_r > 0$,

$$\frac{1}{2\pi} \int_0^{2\pi} |P_r(t)|\,dt = \frac{1}{2\pi} \int_0^{2\pi} P_r(t)\,dt$$

$$= \sum_{j=-\infty}^{\infty} \frac{1}{2\pi} \int_0^{2\pi} r^{|j|} e^{ijt} \, dt$$

$$= \frac{1}{2\pi} \int_0^{2\pi} r^0 e^{i0t} \, dt = 1.$$

This takes care of (1.4.1) and (1.4.2). For (1.4.3), notice that (see Figure 1)

$$|1 - 2r\cos t + r^2| = (r - \cos t)^2 + (1 - \cos^2 t)$$

$$\geq 1 - \cos^2 t$$

$$= \sin^2 t$$

$$\geq \left(\frac{2}{\pi} t\right)^2$$

$$\geq \frac{4}{\pi^2} \epsilon^2$$

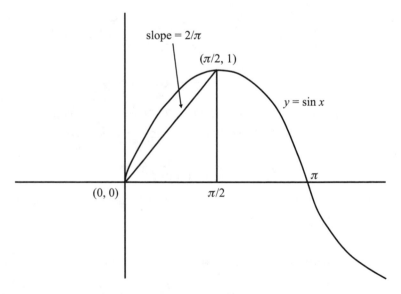

slope = $2/\pi$

$(\pi/2, 1)$

$y = \sin x$

$(0, 0)$ $\pi/2$ π

Figure 1. Estimating the Poisson kernel.

if $\pi/2 \geq |t| \geq \epsilon > 0$. [The estimate for $\pi \geq |t| > \pi/2$ is even easier.] Thus, for such t,

$$|P_r(t)| \leq \frac{\pi^2}{4} \cdot \frac{1 - r^2}{\epsilon^2} \to 0$$

as $r \to 1^-$. Thus the Poisson kernels form a standard family of summability kernels.

Now let us prove a rather general theorem about convergence-inducing properties of families of summability kernels:

Theorem 1.4.4 *Suppose that $\{k_N\}$ is a standard family of summability kernels. If f is any continuous function on $\mathbb{T}$, then*

$$\lim_{N \to \infty} \frac{1}{2\pi} \int_0^{2\pi} f(x - t)k_N(t)\, dt = f(x),$$

with the limit existing uniformly for $x \in \mathbb{T}$.

Proof. Let $\epsilon > 0$. Since f is a continuous function on the compact set $\mathbb{T}$, it is uniformly continuous. Choose $\delta > 0$ such that if $|t| < \delta$ then $|f(x) - f(x - t)| < \epsilon$ for all $x \in \mathbb{T}$. Further, let M be the maximum value of $|f|$ on $\mathbb{T}$.

Using Property (1.4.1) of a standard family, we write

$$\left| \frac{1}{2\pi} \int_0^{2\pi} f(x - t)k_N(t)\, dt - f(x) \right|$$

$$= \left| \frac{1}{2\pi} \int_0^{2\pi} f(x - t)k_N(t)\, dt - \frac{1}{2\pi} \int_0^{2\pi} f(x)k_N(t)\, dt \right|$$

$$= \left| \frac{1}{2\pi} \int_0^{2\pi} [f(x - t) - f(x)]k_N(t)\, dt \right|$$

$$\leq \frac{1}{2\pi} \int_{-\pi}^{\pi} |f(x - t) - f(x)||k_N(t)|\, dt$$

$$= \frac{1}{2\pi} \int_{\{t:|t|<\delta\}} + \frac{1}{2\pi} \int_{\{t:|t|\geq\delta\}}$$

$$\equiv I + II.$$

We notice, by Property (1.4.2) and the choice of δ, that

$$I \leq \frac{1}{2\pi} \int_{-\delta}^{\delta} \epsilon \cdot |k_N(t)| \, dt \leq \epsilon \int_{-\pi}^{\pi} |k_N(t)| \, dt \leq \epsilon \cdot C.$$

That takes care of I.

For II we use Property (1.4.3). If N is large enough, we see that

$$II \leq \frac{1}{2\pi} \int_{\{t:|t|\geq\delta\}} 2M \cdot \epsilon \, dt \leq 2M\epsilon.$$

In summary, if N is sufficiently large, then

$$\left| \frac{1}{2\pi} \int_0^{2\pi} f(x-t)k_N(t) \, dt - f(x) \right| \leq [C + 2M] \cdot \epsilon$$

for all $x \in \mathbb{T}$. This is what we wished to prove. $\square$

We see that, if we use a summability method such as Cesàro's or Abel's to assimilate the Fourier data of a function f, then we may recover *any* continuous function f as the uniform limit of the trigonometric "sums" coming from the Fourier coefficients of f. Contrast this last theorem—especially its proof—with the situation for the ordinary partial sums of the Fourier series. Note that it is Property (1.4.2) that fails for the kernels D_N—see the beginning of this section—and, in fact, makes the theorem false for partial summation. [Property (1.4.3) fails as well, but it is (1.4.2) that will be the focus of our attention.]

It is a general principle in real analysis that if something fails, then it usually fails in a big way. In particular, there is *not* just one continuous function which fails to be the limit of the partial sums of its Fourier series; instead, Baire category theorem arguments (see [RUD2]) can be used to show that the set of functions for which this failure occurs is dense in the space of all continuous functions.

The point of the last paragraph is worth belaboring. For each N, let $\phi_N(t)$ be the function that equals $+1$ when $D_N(t) \geq 0$ and equals -1 when $D_N(t) < 0$. Of course ϕ_N is discontinuous. But now let $\psi_N(t)$ be a continuous function, bounded in absolute value by 1, which agrees with ϕ_N except in a very small interval about each point where ϕ_N changes sign. Integrate D_N against ψ_N. The calculation alluded to at the start of the section then shows that the value of the integral is about $c \cdot \log N$, even though ψ_N has supremum norm 1. The uniform boundedness principle now tells us that convergence for partial summation in norm fails dramatically for continuous functions on the circle group.

The next lemma, due to I. Schur, is key to a number of our elementary norm convergence results:

Lemma 1.4.5 (Schur's Lemma) *Let X and Y be measure spaces equipped with the measures μ and ν, respectively. Let $K(x, y)$ be a measurable kernel on $X \times Y$. Assume that there is a finite constant $M > 0$ such that, for almost every x,*

$$\int_{y \in Y} |K(x, y)| \, d\nu(y) \leq M$$

and, for almost every y,

$$\int_{x \in X} |K(x, y)| \, d\mu(x) \leq M.$$

Then the operator

$$T : f \longmapsto \int_{y \in Y} K(x, y) f(y) \, d\nu(y)$$

is bounded from L^p to L^p, $1 \leq p \leq \infty$. Moreover the operator norm does not exceed M.

Proof. For $p = \infty$ the result is immediate; we leave it as an exercise (or the reader may derive it as a limiting case of $p < \infty$, which we now treat). For $p < \infty$, we use Hölder's inequality to estimate

$$|Tf(x)| \le \int_Y |K(x,y)| \cdot |f(y)| \, dv(y)$$

$$= \int_Y |K(x,y)|^{1/p'} \cdot |K(x,y)|^{1/p} |f(y)| \, dv(y)$$

$$\le \left(\int_Y |K(x,y)| \, dv(y) \right)^{1/p'} \cdot \left(\int_Y |K(x,y)| \cdot |f(y)|^p \, dv(y) \right)^{1/p}$$

$$\le M^{1/p'} \left(\int_Y |K(x,y)| \cdot |f(y)|^p \, dv(y) \right)^{1/p}.$$

We use the last estimate to determine the size of $\|Tf\|_{L^p}$:

$$\|Tf\|_{L^p} \le M^{1/p'} \left(\int_{x \in X} \int_{y \in Y} |K(x,y)| \cdot |f(y)|^p \, dv(y) d\mu(x) \right)^{1/p}$$

$$= M^{1/p'} \left(\int_{y \in Y} \int_{x \in X} |K(x,y)| \, d\mu(x) \, |f(y)|^p \, dv(y) \right)^{1/p}$$

$$\le M^{1/p'+1/p} \left(\int_Y |f(y)|^p \, dv(y) \right)^{1/p}$$

$$= M \cdot \|f\|_{L^p}.$$

That completes the proof. $\square$

Lemma 1.4.5 gives us a straightforward device for applying Functional Analysis Principle I (FAPI). If our operators are given by integration against kernels—$T_j f(x) = \int k_j(x,y) f(y) \, dy$—then, in order to confirm property (0.5.2), it suffices for us to check that $\int_{x \in \mathbb{T}} |k_j(x,y)| \, dx \le C$ and $\int_{y \in \mathbb{T}} |k_j(x,y)| \, dy \le C$.

Now we turn to the topic of "norm convergence" of the summability methods. The question is this: Let $\{k_N\}$ be a family of kernels. Fix $1 \le p \le \infty$. Is it true that, for all $f \in L^p$, we have

$$\lim_{N \to \infty} \left\| \left[\frac{1}{2\pi} \int_0^{2\pi} k_N(t) f(\cdot - t) \, dt \right] - f(\cdot) \right\|_{L^p} = 0 ?$$

This is a question about recovering f "in the mean", rather than point-wise, from the Fourier data. In fact we have the following theorem.

Theorem 1.4.6 Let $1 \le p < \infty$. Let $\{k_N\}$ be a standard family of summability kernels. If $f \in L^p$, then

$$\lim_{N \to \infty} \left\| \frac{1}{2\pi} \int_0^{2\pi} k_N(t) f(\cdot - t) \, dt - f(\cdot) \right\|_{L^p} = 0.$$

Remark: We shall present two proofs of this fundamental result. The first is traditional, and relies on Appendix I. The second is more modern, and uses FAPI.

First Proof of 1.4.6. Fix $f \in L^p$. Let $\epsilon > 0$. Choose $\delta > 0$ so small that if $|s| < \delta$, then $\| f(\cdot - s) - f(\cdot) \|_{L^p} < \epsilon$ (Appendix I). Then

$$\left\| \left[\frac{1}{2\pi} \int_0^{2\pi} k_N(t) f(\cdot - t) \, dt \right] - f(\cdot) \right\|_{L^p}$$

$$= \left[\frac{1}{2\pi} \int_0^{2\pi} \left| \frac{1}{2\pi} \int_0^{2\pi} k_N(t) [f(x - t) \, dt - f(x)] \, dt \right|^p dx \right]^{1/p}$$

$$\le \left[\frac{1}{2\pi} \int_0^{2\pi} \left(\frac{1}{2\pi} \int_0^{2\pi} |k_N(t)[f(x - t) - f(x)]| \, dt \right)^p dx \right]^{1/p}$$

$$\overset{\text{(Minkowski)}}{\le} \frac{1}{2\pi} \int_0^{2\pi} \left[\frac{1}{2\pi} \int_0^{2\pi} |f(x - t) - f(x)|^p \, dx \right]^{1/p} |k_N(t)| \, dt$$

$$= \frac{1}{2\pi} \int_{-\pi}^{\pi} |k_N(t)| \| f(\cdot - t) - f(\cdot) \|_{L^p} \, dt$$

$$= \frac{1}{2\pi} \int_{|t| < \delta} + \frac{1}{2\pi} \int_{|t| \ge \delta}$$

$$\equiv I + II.$$

Now

$$I \leq \frac{1}{2\pi} \int_{|t|<\delta} |k_N(t)| \epsilon \, dt \leq C \cdot \epsilon.$$

For II, we know that if N is sufficiently large, then $|k_N|$ is uniformly small (less than ϵ) on $\{t : |t| \geq \delta\}$. Moreover, we have the easy estimate $\|f(\cdot - t) - f(\cdot)\|_{L^p} \leq 2\|f\|_{L^p}$. Thus

$$II \leq \int_{|t|\geq\delta} \epsilon \cdot 2\|f\|_{L^p} \, dt.$$

This last does not exceed $C' \cdot \epsilon$.

In summary, for all sufficiently large N,

$$\left\| \frac{1}{2\pi} \int_0^{2\pi} k_N(t) f(\cdot - t) \, dt - f(\cdot) \right\|_{L^p} < C'' \cdot \epsilon.$$

That is what we wished to prove.   □

Second Proof of 1.4.6. We know from Theorem 1.4.4 that the desired conclusion is true for continuous functions, and these are certainly dense in L^p (Appendix I). Secondly, we know from Schur's lemma (Lemma 1.4.5) that the operators

$$T_N : f \longmapsto k_N * f$$

are bounded from L^p to L^p with norms bounded by a constant C, independent of N (here we use property (1.4.2) of a standard family of summability kernels). Now FAPI tells us that the conclusion of the theorem follows.   □

Theorem 1.4.6 tells us in particular that the Fejér means of an L^p function f, $1 \leq p < \infty$, converge in norm back to f. It also tells us that the Poisson means converge to f for the same range of p. Finally, Theorem 1.4.4 says that both the Cesàro and the Abel means of a continuous function g converge uniformly to g.

1.5   Pointwise Convergence for Fourier Series

To repeat a point that has been made previously, pointwise convergence for ordinary partial summation of L^p functions, $1 < p < \infty$, is probably the deepest result in all of Fourier analysis. There are two proofs—one due to Carleson-Hunt ([CAR], [HUN]) and the other due to C. Fefferman ([FEF])—and both far exceed the scope of this book.

Pointwise convergence for the standard summability methods is by no means trivial, but it is certainly something that we can discuss here. We do so by way of the Hardy-Littlewood maximal function, an important tool in classical analysis.

Definition 1.5.1 For f an integrable function on $\mathbb{T}$, $x \in \mathbb{T}$, we set

$$Mf(x) = \sup_{R>0} \frac{1}{2R} \int_{x-R}^{x+R} |f(t)| \, dt.$$

The operator M is called the *Hardy-Littlewood maximal operator*.

One can see that Mf is measurable by using the following reasoning. The definition of Mf does not change if $\sup_{R>0}$ is replaced by $\sup_{R>0,\ R \text{ rational}}$. But each average is measurable, and the supremum of countably many measurable functions is measurable. [In fact one can reason a bit differently as follows: Each average is continuous, and the supremum of continuous functions is lower semicontinuous—see [RUD2].]

A priori, it is not even clear that Mf will be finite almost everywhere. We will show that in fact it is, and furthermore obtain an estimate of its relative size.

Lemma 1.5.2 *Let K be a compact set in $\mathbb{T}$. Let $\{U_\alpha\}_{\alpha \in A}$ be a covering of K by open intervals. Then there is a finite subcollection $\{U_{\alpha_j}\}_{j=1}^M$ with the properties:*

(1.5.2.1) *The intervals U_{α_j} are pairwise disjoint.*

(1.5.2.2) *If we take $3U_\alpha$ to be the interval with the same center as U_α but with three times the length, then $\cup_j 3U_{\alpha_j} \supseteq K$.*

Proof. Since K is compact, we may suppose at the outset that the original open cover is finite. So let us call it $\{U_\ell\}_{\ell=1}^p$. Now let U_{ℓ_1} be the open interval among these that has greatest length; if there are several of these longest intervals, then choose one of them. Let U_{ℓ_2} be the open interval, chosen from those remaining, that is disjoint from U_{ℓ_1} and has greatest length—again choose just one if there are several. Continue in this fashion. The process must cease, since we began with only finitely many intervals.

This subcollection does the job. The subcollection chosen is pairwise disjoint by design. To see that the threefold dilates cover K, it is enough to see that the threefold dilates cover the original open cover $\{U_\ell\}_{\ell=1}^p$. Now let U_i be some element of the original open cover. If it is in fact one of the selected intervals, then of course it is covered. If it is *not* one of the selected intervals, then let U_{ℓ_k} be the *first* in the list of selected intervals that intersects U_i (by the selection process, one such must exist). Then, by design, U_{ℓ_k} is at least as long as U_i. Thus, by the triangle inequality, the threefold dilate of U_{ℓ_k} will certainly cover U_i. That is what we wished to prove. $\qquad\square$

Now we may present our boundedness statement for the Hardy-Littlewood maximal function:

Proposition 1.5.3 *If f is an integrable function on $\mathbb{T}$ then, for any $\lambda > 0$,*

$$m\{x \in \mathbb{T} : Mf(x) > \lambda\} \leq \frac{6\pi \|f\|_{L^1}}{\lambda}.$$

Here m stands for the Lebesgue measure of the indicated set. [The displayed estimate is called a weak-type bound for the operator M. See Appendix III for a discursive discussion of the concept of weak-type.]

Proof. By the inner regularity of the measure, it is enough to estimate $m(K)$, where K is any compact subset of $\{x \in \mathbb{T} : Mf(x) > \lambda\}$. Fix such a K, and let $k \in K$. Then, by definition, there is an open interval I_k centered at k such that

$$\frac{1}{m(I_k)} \int_{I_k} |f(t)| \, dt > \lambda.$$

It is useful to rewrite this as

$$m(I_k) < \frac{1}{\lambda} \int_{I_k} |f(t)| \, dt.$$

Now the intervals $\{I_k\}_{k \in K}$ certainly cover K. By the lemma, we may extract a pairwise disjoint subcollection $\{I_{k_j}\}_{j=1}^M$ whose threefold dilates $3 I_{k_j}$ cover K. Putting these ideas together, we see that

$$m(K) \le m \left[\bigcup_{j=1}^M 3 I_{k_j} \right]$$

$$\le 3 \sum_{j=1}^M m(I_{k_j})$$

$$\le 3 \sum_{j=1}^M \frac{1}{\lambda} \int_{I_{k_j}} |f(t)| \, dt.$$

Note that the intervals over which we are integrating in the last sum on the right are pairwise disjoint. So we may majorize the sum of these integrals by the L^1 norm of f. In other words,

$$m(K) \le \frac{3 \cdot 2\pi}{\lambda} \|f\|_{L^1}.$$

This is what we wished to prove. $\qquad\qquad\qquad\qquad\qquad\qquad\qquad\square$

Of course the maximal operator is trivially bounded on L^∞. It then follows from the Marcinkiewicz interpolation theorem (Appendix III) that M is bounded on L^p for $1 < p \le \infty$.

The maximal operator is certainly unbounded on L^1. To see this, calculate the maximal function of

$$f_N(x) = \begin{cases} N & \text{if } |x| \leq \frac{1}{2N} \\ 0 & \text{if } |x| > \frac{1}{2N}. \end{cases}$$

Each of the functions f_N has L^1 norm 1, but their associated maximal functions have L^1 norms that blow up. [Remember that we are working on the circle group $\mathbb{T}$; the argument is even easier on the real line, for Mf_1 is already not integrable at infinity.]

And now the key fact is that, in an appropriate sense, each of our families of standard summability kernels is majorized by the Hardy-Littlewood maximal function. This assertion must be checked in detail, and by a separate argument, for each particular family of summability kernels. To illustrate the ideas, we will treat the Poisson family at this time.

Proposition 1.5.4 *There is a constant finite $C > 0$ such that, if $f \in L^1(\mathbb{T})$, then*

$$P^* f(e^{i\theta}) \equiv \sup_{0 < r < 1} |P_r f(e^{i\theta})| \leq C M f(e^{i\theta})$$

for all $\theta \in [0, 2\pi)$.

Proof. The estimate for $0 < r \leq 1/2$ is easy (do it as an exercise— either using the maximum principle or by imitating the proof that we now present). Thus we concentrate on $1/2 < r < 1$. We estimate

$$|P_r f(e^{i\theta})| = \left| \frac{1}{2\pi} \int_0^{2\pi} f(e^{i(\theta - \psi)}) \frac{1 - r^2}{1 - 2r \cos \psi + r^2} \, d\psi \right|$$

$$= \left| \frac{1}{2\pi} \int_{-\pi}^{\pi} f(e^{i(\theta - \psi)}) \frac{1 - r^2}{(1 - r)^2 + 2r(1 - \cos \psi)} \, d\psi \right|$$

$$\leq \frac{1}{2\pi} \sum_{j=0}^{\log_2(\frac{\pi}{1-r})} \int_{S_j} |f(e^{i(\theta-\psi)})| \frac{1-r^2}{(1-r)^2 + 2r(2^{j-2}(1-r))^2} \, d\psi$$

$$+ \frac{1}{2\pi} \int_{|\psi|<1-r} |f(e^{i(\theta-\psi)})| \frac{1-r^2}{(1-r)^2} \, d\psi,$$

where $S_j = \{\psi : 2^j(1-r) \leq |\psi| < 2^{j+1}(1-r)\}$. Now this last expression is (since $1 + r < 2$)

$$\leq \frac{1}{\pi} \sum_{j=0}^{\infty} \frac{1}{2^{2j-4}(1-r)} \int_{|\psi|<2^{j+1}(1-r)} |f(e^{i(\theta-\psi)})| \, d\psi$$

$$+ \frac{1}{\pi} \frac{1}{1-r} \int_{|\psi|<1-r} |f(e^{i(\theta-\psi)})| \, d\psi$$

$$\leq \frac{64}{\pi} \sum_{j=0}^{\infty} 2^{-j} \left[\frac{1}{2 \cdot 2^{j+1}(1-r)} \int_{|\psi|<2^{j+1}(1-r)} |f(e^{i(\theta-\psi)})| \, d\psi \right]$$

$$+ \frac{2}{\pi} \left[\frac{1}{2(1-r)} \int_{|\psi|<1-r} |f(e^{i(\theta-\psi)})| \, d\psi \right]$$

$$\leq \frac{64}{\pi} \cdot \sum_{j=0}^{\infty} 2^{-j} Mf(e^{i\theta}) + \frac{2}{\pi} Mf(e^{i\theta})$$

$$\leq \frac{128}{\pi} Mf(e^{i\theta}) + \frac{2}{\pi} Mf(e^{i\theta})$$

$$\equiv C \cdot Mf(e^{i\theta}).$$

This is the desired estimate. □

Corollary 1.5.5 *The operator* $P^*f \equiv \sup_{0<r<1} P_r f$ *is weak-type* $(1, 1)$.

Proof. We know that P^*f is majorized by a constant times the Hardy-Littlewood maximal operator of f. Since (by 1.5.3) the latter is weak-type $(1, 1)$, the result follows. □

Now we will invoke our Second Functional Analysis Principle from Chapter 1 to derive a pointwise convergence result.

Theorem 1.5.6 *Let f be an integrable function on $\mathbb{T}$. Then, for almost every $x \in \mathbb{T}$, we have that*

$$\lim_{r \to 1^-} \frac{1}{2\pi} \int_0^{2\pi} f(t) P_r(x - t) \, dt = f(x).$$

Proof. The remarkable thing to notice is that this result now follows with virtually no additional work: First observe that the continuous functions are dense in $L^1(\mathbb{T})$. Secondly, Theorem 1.4.4 tells us that the desired conclusion holds for functions in this dense set. Finally, Corollary 1.5.5 gives that P^* is weak-type $(1, 1)$. That is the key ingredient that enables us to apply FAPII. The result follows: Poisson summation of Fourier series is valid in L^1. □

The reader should take special note that this theorem, whose proof appears to be a rather abstract manipulation of operators, says that a fairly "arbitrary" function f may be recovered, pointwise almost everywhere, by the Poisson summation method from the Fourier data of f.

A similar theorem holds for the Fejér summation method. We invite the interested reader to show that the maximal Fejér means of an L^1 function are bounded above (pointwise) by a multiple of the Hardy-Littlewood maximal function. The result is then automatic from the machinery that we have set up.

1.6 Norm Convergence of Partial Sums and the Hilbert Transform

In this section we discontinue our discussion of summation methods and return to the more basic consideration of ordinary partial sums.

Naively, pointwise convergence of Fourier series, or of any series of functions, would seem to be the ultimate goal of any study. But in fact many applications require what appears to be a somewhat coarser notion of convergence: norm convergence. The present section is devoted to that idea.

While many of the ideas presented here may be applied to *almost any* Banach space norm, we shall in fact confine our attention to L^p norms. The fundamental question is this:

If $f \in L^p(\mathbb{T})$, $1 \le p \le \infty$, then is it the case that $S_N f \to f$ in the L^p topology? Precisely, is it true that

$$\left[\int_0^{2\pi} |S_N f(e^{it}) - f(e^{it})|^p \, dt \right]^{1/p} \to 0$$

as $N \to \infty$?

Of course there are many variants of the question. In Section 1.4, we answered this question with S_N replaced by σ_N (Fejér summation) or P_r (Poisson summation). As with pointwise convergence, the use of these summation methods makes the study of norm convergence questions much easier.

We have already noted that the Dirichlet kernels D_N satisfy

$$\|D_N\|_{L^1} \approx c \cdot \log N.$$

In particular, these kernels do *not* satisfy the hypothesis of Schur's lemma uniformly. So the Dirichlet kernels do not form a standard family of kernels. This is a fundamental and ineluctable fact about Fourier analysis. It means that *if* norm convergence is going to work for partial summation—the most simple and natural method for adding up the terms of a Fourier series—then it will have to be for a more subtle reason.

In fact norm-convergent partial summation in L^p is valid for $1 < p < \infty$. It is false for $p = 1, \infty$. We shall treat both these facts, in some detail, in this section. The reader will find, as the book progresses, that we frequently return to these fundamental ideas and cast new light on them using ideas that we develop along the way.

We begin with the theory of L^2 convergence. This is an exercise in "soft" analysis,[2] for it consists of interpreting some elementary Hilbert space ideas (Section 0.4) for the particular Hilbert space $L^2(\mathbb{T})$. As usual, we define

$$L^2(\mathbb{T}) \equiv \left\{ f \text{ measurable on } \mathbb{T} : \|f\|_{L^2} \right.$$

$$\left. \equiv \left[\frac{1}{2\pi} \int_0^{2\pi} |f(e^{it})|^2 \, dt \right]^{1/2} < \infty \right\}.$$

As already noted in Chapter 0, L^2 is equipped with the inner product

$$\langle f, g \rangle = \frac{1}{2\pi} \int f(e^{it}) \overline{g(e^{it})} \, dt$$

and the induced metric

$$\mathbf{d}(f, g) = \|f - g\|_{L^2} = \sqrt{\frac{1}{2\pi} \int |f(e^{it}) - g(e^{it})|^2 \, dt}$$

(under which L^2 is complete).

Observe that the sequence of functions $\mathcal{F} \equiv \{e^{ijt}\}_{j=-\infty}^{\infty}$ is a complete orthonormal basis for L^2. [It will sometimes be useful to write $e_j(t) = e^{ijt}$.] The orthonormality is obvious, and the completeness can be seen by noting that the algebra generated by the exponential functions satisfies the hypotheses of the Stone-Weierstrass theorem (Appendix VIII) on the circle group $\mathbb{T}$. Thus the trigonometric polynomials are uniformly dense in $C(\mathbb{T})$, the continuous functions on $\mathbb{T}$. If f is an L^2 function that is orthogonal to every e^{ijx}, then it is orthogonal to all trigonometric polynomials and hence to all continuous functions on $\mathbb{T}$. But the continuous functions are dense in L^2. So it must be that $f \equiv 0$ and the family $\mathcal{F}$ is complete. [See Appendix I for a discussion of related ideas.] The fact that the group characters for the circle group $\mathbb{T}$ also form a complete orthonormal system for the Hilbert space $L^2(\mathbb{T})$

---

[2] Analysts call an argument "soft" if it does not use estimates, particularly if it does not use ϵ's and δ's.

(which is a special case of the Peter-Weyl theorem—see [FOL2]) will play a crucial role in what follows.

In fact we shall treat the quadratic theory of Fourier integrals in considerable detail in Chapter 2—specifically Section 2.3. We shall take this opportunity to summarize some of the key facts about Fourier series of L^2 functions. Their proofs are transcriptions either of general facts about Hilbert space or of particular arguments presented in Chapter 2 for the Fourier integral.

Proposition 1.6.1 (Bessel's Inequality) *Let* $f \in L^2(\mathbb{T})$. *Let* N *be a positive integer and let* $a_j = \widehat{f}(j)$, *each* j. *Then*

$$\sum_{j=-N}^{N} |a_j|^2 \leq \|f\|_{L^2}^2.$$

Theorem 1.6.2 (Riesz-Fischer) *Let* $\{a_j\}_{j=-\infty}^{\infty}$ *be a square-summable sequence (i.e.,* $\sum_j |a_j|^2$ *is finite). Then the series*

$$\sum_{j=-\infty}^{\infty} a_j e^{ijx}$$

converges in the L^2 *topology. It defines a function* $f \in L^2(\mathbb{T})$. *Moreover, for each* n,

$$\widehat{f}(n) = a_n.$$

Theorem 1.6.3 (Parseval's Formula) *Let* $f \in L^2(\mathbb{T})$. *Then the sequence* $\{\widehat{f}(j)\}$ *is square-summable and*

$$\frac{1}{2\pi} \int_0^{2\pi} |f(x)|^2 \, dx = \sum_{j=-\infty}^{\infty} |\widehat{f}(j)|^2.$$

Exercise. Apply Parseval's formula to the function $f(x) = x$ on the interval $[0, 2\pi]$. Actually calculate the integral on the left, and write out the terms of the series on the right. Conclude that

$$\frac{\pi^2}{6} = \sum_{j=1}^{\infty} \frac{1}{j^2}. \qquad \square$$

The next result is the key to our treatment of the L^2 theory of norm convergence. It is also a paradigm for the more elaborate L^p theory that we treat afterward.

Proposition 1.6.4 *Let* $\Lambda = \{\lambda_j\}$ *be a sequence of complex numbers. Then the multiplier operator* $\mathcal{M}_\Lambda$ *is bounded on* L^2 *if and only if* Λ *is a bounded sequence. Moreover, the supremum norm of the sequence is equal to the operator norm of the multiplier operator.*

Proof. This is a direct application of Plancherel's theorem (Theorem 2.3.16). The details are provided in the proof of Proposition 3.3.1. $\square$

Of course the multiplier corresponding to the partial-summation operator S_N is just the sequence $\Lambda^N \equiv \{\lambda_j\}$ given by

$$\lambda_j = \begin{cases} 1 & \text{if} \quad |j| \le N \\ 0 & \text{if} \quad |j| > N. \end{cases}$$

[In what follows, we will often denote this particular multiplier by $\chi_{[-N,N]}$. It should be clearly understood that the domain of $\chi_{[-N,N]}$ is the set of integers $\mathbb{Z}$.] This sequence is bounded by 1, so the proposition tells us that S_N is bounded on L^2. But in fact more is true. We know that $\|S_N\|_{\text{op}} = 1$ for every N. So the operators S_N are uniformly bounded in norm. In addition, the trigonometric polynomials are dense in L^2, and if p is such a polynomial then, when N exceeds the degree of p, $S_N(p) = p$; therefore norm convergence obtains for p. By Functional Analysis Principle I, we conclude that norm convergence is valid in L^2. More precisely:

Theorem 1.6.5 *Let $f \in L^2(\mathbb{T})$. Then $\|S_N f - f\|_{L^2} \to 0$ as $N \to \infty$.* *Explicitly,*

$$\lim_{N \to \infty} \left[\int_{\mathbb{T}} |S_N f(x) - f(x)|^2 \, dx \right]^{1/2} = 0.$$

Next we turn to the Hilbert transform H, which is one of the most important linear operators in all of mathematical analysis. First, it is the key to all convergence questions for the partial sums of Fourier series. Second, it is a paradigm for all singular integral operators on Euclidean space (we shall treat these in Chapter 5). Third, the analogue of the Hilbert transform on the line is uniquely determined by its invariance properties with respect to the groups that act naturally on 1-dimensional Euclidean space.

We begin by defining the Hilbert transform as a multiplier operator. Indeed, let $\mathbf{h} = \{h_j\}$, with $h_j = -i \operatorname{sgn} j$; here the convention is that

$$\operatorname{sgn} x = \begin{cases} -1 & \text{if} \quad x < 0 \\ 0 & \text{if} \quad x = 0 \\ 1 & \text{if} \quad x > 0. \end{cases}$$

Then $H \equiv \mathcal{M}_{\mathbf{h}}$. So defined, the Hilbert transform has the following connection with the partial sum operators:

$$\chi_{[-N,N]}(j) = \frac{1}{2}\big[1 + \operatorname{sgn}(j + N)\big] - \frac{1}{2}\big[1 + \operatorname{sgn}(j - N)\big]$$

$$+ \frac{1}{2}\big[\chi_{\{-N\}}(j) + \chi_{\{N\}}(j)\big]$$

$$= \frac{1}{2}\big[\operatorname{sgn}(j + N) - \operatorname{sgn}(j - N)\big]$$

$$+ \frac{1}{2}\big[\chi_{\{-N\}}(j) + \chi_{\{N\}}(j)\big].$$

See Figure 1. Therefore

$$S_N f(e^{it}) = \frac{1}{2} i e^{-iNt} H[e_N f] - \frac{1}{2} i e^{iNt} H[e_{-N} f]$$

$$+ \frac{1}{2} \left[P_{-N} f + P_N f \right], \tag{1.6.6}$$

where P_j is orthogonal projection onto the space spanned by e_j.

To understand this last equality, let us examine a piece of it. Let us look at the linear operator corresponding to the multiplier

$$m(j) \equiv \text{sgn}\,(j + N).$$

Let $f(t) \sim \sum_{j=-\infty}^{\infty} \widehat{f}(j) e^{itj}$. Then

$$\mathcal{M}_m f(t) = \sum_j \text{sgn}\,(j + N) \widehat{f}(j) e^{ijt}$$

$$= \sum_j \text{sgn}\,(j) e^{-iNt} \widehat{f}(j - N) e^{ijt}$$

$$= i e^{-iNt} \sum_j (-i) \text{sgn}\,(j) \widehat{f}(j - N) e^{ijt}$$

$$= i e^{-iNt} \sum_j (-i) \text{sgn}\,(j) \big(e_N f\big)^{\widehat{}}(j) e^{ijt}$$

$$= i e^{-iNt} H[e_N f](t).$$

This is of course precisely what is asserted in the first half of the right-hand side of (1.6.6).

We know that the Hilbert transform is bounded on L^2 because it is a multiplier operator coming from a bounded sequence. It also turns out to be bounded on L^p for $1 < p < \infty$. [We shall discuss this fact about H below, and eventually prove it in Chapter 6.] Similar remarks apply to the projection operators P_j. Taking these boundedness assertions for granted, we now reexamine equation (1.6.6). Multiplication by a complex exponential does not change the size of an L^p function (in technical language, it is an *isometry* of L^p). So (1.6.6) tells us that S_N is a difference of compositions of operators, all of which are bounded on

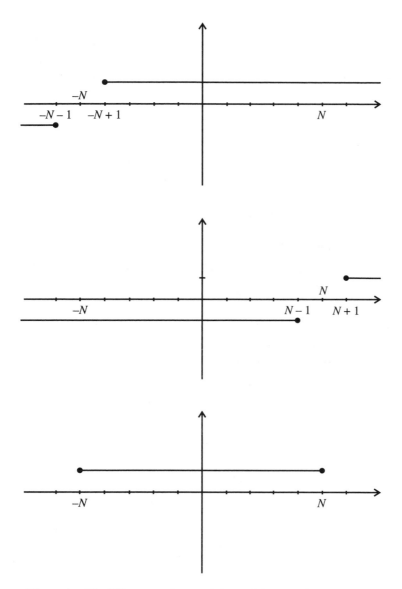

Figure 1.   The Hilbert transform and the partial summation operators.

L^p. And the norm is plainly bounded independent of N. In conclusion, if we assume that H is bounded on L^p, $1 < p < \infty$, then Functional Analysis Principle I tells us (since trigonometric polynomials are dense in L^p for $1 \le p < \infty$) that norm convergence holds in L^p for $1 < p < \infty$. We now state this as a theorem:

Theorem 1.6.7 *Let* $1 < p < \infty$ *and* $f \in L^p(\mathbb{T})$. *Then* $\|S_N f - f\|_{L^p} \to 0$ *as* $N \to \infty$. *Explicitly,*

$$\lim_{N \to \infty} \left[\int_{\mathbb{T}} |S_N f(x) - f(x)|^p \, dx \right]^{1/p} = 0.$$

It is useful in the study of the Hilbert transform to be able to express it explicitly as an integral operator. The next lemma is of great utility in this regard.

Lemma 1.6.8 *If the Fourier multiplier* $\Lambda = \{\lambda_j\}_{j=-\infty}^{\infty}$ *induces a bounded operator* $\mathcal{M}_\Lambda$ *on* L^p, *then the operator is given by a convolution kernel* K. *In other words,*

$$\mathcal{M}_\Lambda f(x) = f * K(x) = \frac{1}{2\pi} \int_0^{2\pi} f(t) K(x - t) \, dt.$$

That convolution kernel is specified by the formula

$$K(e^{it}) = \sum_{j=-\infty}^{\infty} \lambda_j e^{it}.$$

[In practice, the sum that defines this kernel may have to be interpreted using a summability technique, or using distribution theory, or both.]

Proof. A rigorous proof of this lemma would involve a digression into distribution theory and the Schwartz kernel theorem. We refer the interested reader to either [STG1] or [SCH]. $\square$

In practice, when we need to get our hands on the kernel associated to a multiplier, we will just calculate it. So this lemma will play a moot role in our work.

If we apply Lemma 1.6.8 directly to the multiplier for the Hilbert transform, we obtain the formal series

$$\sum_{j=-\infty}^{\infty} -i \cdot \operatorname{sgn} j \cdot e^{ijt}.$$

Of course the terms of this series do not tend to zero, so this series does not converge in any conventional sense. Instead we use Abel summation (i.e., summation with factors of $r^{|j|}$, $0 \le r < 1$) to interpret the series: For $0 \le r < 1$ let

$$k_r(e^{it}) = \sum_{j=-\infty}^{\infty} -ir^{|j|} \cdot \operatorname{sgn} j \cdot e^{ijt}.$$

The sum over the positive indices is

$$-i \sum_{j=1}^{\infty} r^j \cdot e^{ijt} = -i \sum_{j=1}^{\infty} [re^{it}]^j$$

$$= -i \left[\frac{1}{1 - re^{it}} - 1 \right]$$

$$= \frac{-ire^{it}}{1 - re^{it}}.$$

Similarly, the sum over negative indices can be calculated to be equal to

$$\frac{ire^{-it}}{1 - re^{-it}}.$$

Adding these two pieces yields that

$$k_r(e^{it}) = \frac{-ire^{it}}{1 - re^{it}} + \frac{ire^{-it}}{1 - re^{-it}}$$

$$= \frac{-ir[e^{it} - e^{-it}]}{|1 - re^{it}|^2}$$

$$= \frac{2r \sin t}{|1 - re^{it}|^2}$$

$$= \frac{2r \sin t}{1 + r^2 - 2r \cos t}$$

$$= \frac{2r \cdot 2 \cdot \sin \frac{t}{2} \cos \frac{t}{2}}{(1 + r^2 - 2r) + 2r(1 - \cos^2 \frac{t}{2} + \sin^2 \frac{t}{2})}$$

$$= \frac{4r \sin \frac{t}{2} \cos \frac{t}{2}}{(1 + r^2 - 2r) + 2r(2 \sin^2 \frac{t}{2})}.$$

We formally let $r \to 1^-$ to obtain the kernel

$$k(e^{it}) = \frac{\sin \frac{t}{2} \cos \frac{t}{2}}{\sin^2 \frac{t}{2}} = \cot \frac{t}{2}. \tag{1.6.9}$$

This is the standard formula for the kernel of the Hilbert transform. It should be noted that we suppressed various subtleties about the validity of Abel summation in this context, as well as issues concerning the fact that the kernel k is not integrable. For the full story, consult [KAT].

We resolve the nonintegrability problem for the integral kernel k in (1.6.9) by using the so-called *Cauchy principal value*; the Cauchy principal value is denoted P.V. and will now be defined. Thus we usually write

$$\text{P.V.} \frac{1}{2\pi} \int_{-\pi}^{\pi} f(x - t) \cot \left(\frac{t}{2} \right) dt,$$

and we interpret this to mean

$$\lim_{\epsilon \to 0^+} \frac{1}{2\pi} \int_{\epsilon < |t| \leq \pi} f(x - t) \cot \left(\frac{t}{2} \right) dt. \tag{1.6.10}$$

Observe in (1.6.10) that, for $\epsilon > 0$ fixed, $\cot(t/2)$ is actually *bounded* on the domain of integration. Therefore the integral in (1.6.10) makes sense, by Hölder's inequality, as long as $f \in L^p$ for some $1 \leq p \leq \infty$.

The deep question is whether the limit exists, and whether that limit defines an L^p function.

We now reiterate the most fundamental fact about the Hilbert transform in the language of integral operators:

Theorem 1.6.11 (M. Riesz) *The operator*

$$Hf(x) \equiv \text{P.V.} \ \frac{1}{2\pi} \int_{-\pi}^{\pi} f(x-t) \cot\left(\frac{t}{2}\right) dt$$

is defined on $L^p(\mathbb{T})$, $1 \leq p \leq \infty$. It is bounded on L^p when $1 < p < \infty$, but is unbounded on L^1 and L^∞.

We will prove Theorem 1.6.8, after some considerable additional development, in Chapter 6. The significance of Theorem 1.6.11 is given by Theorem 1.6.12:

Theorem 1.6.12 *Let $1 \leq p \leq \infty$. Norm-convergent partial summation of Fourier series is valid in the L^p topology if and only if the integral operator in (1.6.10) is bounded on L^p.*

An immediate corollary of these two theorems taken together is that norm-convergent partial summation of Fourier series is valid in L^p if and only if $1 < p < \infty$.

The result just enunciated is fundamental to the study of Fourier series. But it also holds great philosophical significance in the modern history of analysis. For it shows that we may reduce the study of the (infinitely many) partial sums of the Fourier series of a function to the study of a *single* integral operator. The device for making this reduction is—rather than study one function at a time—to study an entire space of functions at once. Many of the basic ideas in functional analysis—including the uniform boundedness principle, the open mapping theorem, and the Hahn-Banach theorem—grew out of questions of Fourier analysis.

In the next section we shall examine the Hilbert transform from another point of view.

In the present section, we shall take the validity of Theorem 1.6.11 and (most of) Theorem 1.6.12 for granted. Those details will be treated as the book develops. Our intention now is to discuss these theorems, and to look at some examples.

We first note that, in practice, people do not actually look at the operator consisting of convolution with $\cot \frac{t}{2}$. This kernel is a transcendental function, and is tedious to handle. Thus what we do instead is to look at the operator

$$\widetilde{H} : f \longmapsto \text{P.V.} \frac{1}{2\pi} \int_{-\pi}^{\pi} f(x - t) \cdot \frac{2}{t} \, dt. \qquad (1.6.13)$$

Clearly the kernel $2/t$ is much easier to think about than $\cot \frac{t}{2}$. It is also homogeneous of degree -1, a fact that will prove significant when we adopt a broader point of view later. But what gives us the right to replace the complicated integral (1.6.10) by the apparently simpler integral (1.6.13)?

Let us examine the difference

$$I(t) \equiv \cot \left(\frac{t}{2} \right) - \frac{2}{t}, \quad 0 < |t| < \pi,$$

which we extend to $\mathbb{R} \setminus \pi \mathbb{Z}$ by 2π-periodicity. Using the Taylor expansions of sine and cosine near the origin, we may write

$$\begin{aligned}
I(t) &= \frac{1 + \mathcal{O}(t^2)}{t/2 + \mathcal{O}(t^3)} - \frac{2}{t} \\
&= \frac{2 + \mathcal{O}(t^2)}{t + \mathcal{O}(t^3)} - \frac{2}{t} \\
&= \frac{2}{t} \cdot \left[\frac{1 + \mathcal{O}(t^2)}{1 + \mathcal{O}(t^2)} - 1 \right] \\
&= \frac{2}{t} \cdot \mathcal{O}(t^2) \\
&= \mathcal{O}(t).
\end{aligned}$$

Thus the difference between the two kernels under study is (better than) a bounded function. In particular, it is in every L^p class. So we think of

$$\text{P.V.} \frac{1}{2\pi} \int_{-\pi}^{\pi} f(x-t) \cot\left(\frac{t}{2}\right) dt$$

$$= \text{P.V.} \frac{1}{2\pi} \int_{-\pi}^{\pi} f(x-t) \cdot \frac{2}{t} dt$$

$$+ \text{P.V.} \frac{1}{2\pi} \int_{-\pi}^{\pi} f(x-t) \cdot I(t) dt$$

$$= \widetilde{H} f(x) + \frac{1}{2\pi} \int_{-\pi}^{\pi} f(t) I(x-t) dt$$

$$\equiv J_1 + J_2.$$

By Schur's Lemma (Lemma 1.4.5), the integral J_2 is trivial to study: The integral operator

$$\phi \mapsto \phi * I$$

is bounded on every L^p space. Thus, in order to study the Hilbert transform, it suffices for us to study the integral in (1.6.13). In practice, harmonic analysts study the integral in (1.6.13) and refer to it as the "(modified) Hilbert transform" without further comment. To repeat, the beautiful fact is that the original Hilbert transform is bounded on a given L^p space *if and only if* the new transform (1.6.13) is bounded on that same L^p. [In practice it is convenient to forget about the "2" in the numerator of the kernel in (1.6.13).]

Proposition 1.6.14 *Norm summability for Fourier series fails in both L^1 and L^∞.*

Proof. It suffices for us to show that the modified Hilbert transform (as defined in (1.6.13)) fails to be bounded on L^1 and fails to be bounded on L^∞. In fact the following lemma will cut the job by half:

Lemma 1.6.15 *If the modified Hilbert transform $\widetilde{H}$ is bounded on L^1, then it is bounded on L^∞.*

Proof. Let f be an L^∞ function. Then (using the remarks following Theorem 0.3.9)

$$\|\widetilde{H}f\|_{L^\infty} = \sup_{\substack{\phi \in L^1 \\ \|\phi\|_{L^1}=1}} \left| \int \widetilde{H}f(x) \cdot \phi(x)\, dx \right|$$

$$= \sup_{\substack{\phi \in L^1 \\ \|\phi\|_{L^1}=1}} \left| \int f(x)(\widetilde{H}^*\phi)(x)\, dx \right|.$$

But an easy formal argument shows that

$$\widetilde{H}^*\phi = -\widetilde{H}\phi.$$

[In fact a similar formula holds for *any* convolution operator—exercise.] Thus the last line gives

$$\|\widetilde{H}f\|_{L^\infty} = \sup_{\substack{\phi \in L^1 \\ \|\phi\|_{L^1}=1}} \left| \int f(x)\widetilde{H}\phi(x)\, dx \right|$$

$$\leq \sup_{\substack{\phi \in L^1 \\ \|\phi\|_{L^1}=1}} \|f\|_{L^\infty} \cdot \|\widetilde{H}\phi\|_{L^1}$$

$$\leq \sup_{\substack{\phi \in L^1 \\ \|\phi\|_{L^1}=1}} \|f\|_{L^\infty} \cdot C\|\phi\|_{L^1}$$

$$= C \cdot \|f\|_{L^\infty}.$$

Here C is the norm of the modified Hilbert transform acting on L^1. We have shown that if $\widetilde{H}$ is bounded on L^1, then it is bounded on L^∞. That is the claim. We frequently summarize this type of argument by invoking "duality." $\qquad\square$

Remark: In fact the proposition that we just proved is true in considerable generality. We may replace L^1, L^∞ by any two conjugate spaces L^p, $L^{p'}$ with $1/p + 1/p' = 1$. And we may replace the Hilbert transform by any convolution operator. Details are left for the interested reader.

Now let us resume the proof of Proposition 1.6.14. Let $f = \chi_{[0,a]}$, where a is a small positive number and χ denotes the characteristic function of the indicated interval. We may calculate the (modified) Hilbert transform of f by hand and see that

$$\widetilde{H} f(x) = c \cdot \log \frac{|x|}{|x - a|}.$$

In particular, $\widetilde{H} f$ is an *unbounded* function. Therefore the Hilbert transform is not bounded on L^∞. [If we accept Theorems 1.6.11 and 1.6.12, we might have predicted that boundedness of the Hilbert transform would fail on L^∞ by just a logarithm because $\|D_N\|_{L^1} \approx \log N$.] By Lemma 1.6.15, we may also conclude that boundedness of the Hilbert transform fails on L^1.

It follows from the arguments in the last paragraph that norm summability for Fourier series fails in L^1 and L^∞. □

We close this section with a commentary about an interpretation of the Hilbert transform in terms of analytic function theory. Let ϕ be any L^2 function on the circle group $\mathbb{T}$. Then

$$\phi(e^{it}) = \sum_{j=-\infty}^{\infty} a_j e^{ijt},$$

where the a_j are the Fourier coefficients of ϕ. Here the coefficients form a square-summable sequence, and the convergence of the partial sums is in L^2—the details of these assertions may be found in Sections 1.6 and 2.3.

Instead of looking at the Hilbert transform H, let us consider the closely related operator $J = \frac{1}{2}[I + iH]$, where I is the identity. A simple calculation with Fourier multipliers shows that

$$J : \sum_{j=-\infty}^{\infty} a_j e^{ijt} \longmapsto \sum_{j=1}^{\infty} a_j e^{ijt} + \frac{1}{2} a_0.$$

Now we observe that each function e^{ijt} has a unique harmonic extension to the disc: the extension when $j \geq 0$ is just z^j and that when

$j < 0$ is $\bar{z}^{|j|}$. The harmonic function that solves the Dirichlet problem (i.e., the problem of constructing a harmonic function with specified boundary data) with boundary data ϕ is

$$\Phi(z) \equiv \sum_{j=-\infty}^{-1} a_j \bar{z}^{|j|} + \sum_{j=0}^{\infty} a_j z^j.$$

In this language, we see that J assigns to the function Φ its "holomorphic part" $\sum_{j=1}^{\infty} a_j z^j + \frac{1}{2} a_0$.

We leave it as an exercise for you to think about the relationship of this interpretation of the Hilbert transform with the interpretation discussed in the text (Section 5.5) of the Hilbert transform as the operator that assigns to a harmonic function on the disc its harmonic conjugate.

We end this section by recording what is perhaps the deepest result of basic Fourier analysis. Formerly known as the Lusin conjecture, and now as Carleson's theorem, this result addresses the pointwise convergence question for L^2.

Theorem 1.6.16 (Carleson [CAR]) *Let $f \in L^2(\mathbb{T})$. Then the Fourier series of f converges almost everywhere to f.*

The next result is based on Carleson's theorem, but requires significant new ideas.

Theorem 1.6.17 (Hunt [HUN]) *Let $f \in L^p(\mathbb{T})$, $1 < p \leq \infty$. Then the Fourier series of f converges almost everywhere to f.*

P. Sjölin [SJO1] has refined Hunt's theorem even further to obtain spaces of functions that are smaller than L^1, yet larger than L^p for every $p > 1$, on which pointwise convergence of Fourier series holds. The sharpest result along these lines is due to Hunt and Taibleson [HUT].

A classical example of A. Kolmogorov (see [KAT], [ZYG]) provides an L^1 function whose Fourier series converges *at no point* of $\mathbb{T}$.

This phenomenon provides significant information: If instead the example were of a function with Fourier series diverging a.e., then we might conclude that we were simply using the wrong measure. But since there is an L^1 function with everywhere diverging Fourier series, we conclude that there is no hope for pointwise convergence in L^1.

CHAPTER 2

The Fourier Transform

2.1 Basic Properties of the Fourier Transform

A thorough treatment of the Fourier transform in Euclidean space may be found in [STG1]. Here we give a sketch of the theory. Most of the results parallel facts that we have already seen in the context of Fourier series on the circle. But some, such as the invariance properties of the Fourier transform under the groups that act on Euclidean space (Section 2.2), will be new.

If $t, \xi \in \mathbb{R}^N$ then we let

$$t \cdot \xi \equiv t_1 \xi_1 + \cdots + t_N \xi_N.$$

We define the *Fourier transform* of a function $f \in L^1(\mathbb{R}^N)$ by

$$\widehat{f}(\xi) = \int_{\mathbb{R}^N} f(t) e^{it \cdot \xi} \, dt.$$

Here dt denotes Lebesgue N-dimensional measure. Many references will insert a factor of 2π in the exponential or in the measure. Others will insert a minus sign in the exponent. There is no agreement on this matter. We have opted for this definition because of its simplicity.

We note that the significance of the exponentials $e^{it \cdot \xi}$ is that the only continuous multiplicative homomorphisms of $\mathbb{R}^N$ into the circle group are the functions $\phi_\xi(t) = e^{it \cdot \xi}$, $\xi \in \mathbb{R}^N$. [We leave it to the reader to observe that the argument provided in Section 1.1 for the circle group works nearly verbatim here.] These functions are called the *characters* of the additive group $\mathbb{R}^N$.

Proposition 2.1.1 *If $f \in L^1(\mathbb{R}^N)$, then*

$$\|\widehat{f}\|_{L^\infty(\mathbb{R}^N)} \le \|f\|_{L^1(\mathbb{R}^N)}.$$

In other words, $\widehat{}$ is a bounded operator from L^1 to L^∞. We sometimes denote the operator by $\mathcal{F}$.

Proof. Observe that, for any $\xi \in \mathbb{R}^N$,

$$|\widehat{f}(\xi)| \le \int |f(t)|\, dt. \qquad \square$$

Proposition 2.1.2 *If $f \in L^1(\mathbb{R}^N)$, f is differentiable, and $\partial f/\partial x_j \in L^1(\mathbb{R}^N)$, then*

$$\left(\frac{\partial f}{\partial x_j}\right)^{\widehat{}}(\xi) = -i\xi_j \widehat{f}(\xi).$$

Proof. Integrate by parts: If $f \in C_c^\infty$ (see Appendix I), then

$$\left(\frac{\partial f}{\partial x_j}\right)^{\widehat{}}(\xi) = \int \frac{\partial f}{\partial t_j} e^{it\cdot\xi}\, dt$$

$$= \int \cdots \int \left[\int \frac{\partial f}{\partial t_j} e^{it\cdot\xi}\, dt_j\right] dt_1 \ldots dt_{j-1} dt_{j+1} \ldots dt_N$$

$$= -\int \cdots \int f(t) \left(\frac{\partial}{\partial t_j} e^{it\cdot\xi}\right) dt_j dt_1 \ldots dt_{j-1} dt_{j+1} \ldots dt_N$$

$$= -i\xi_j \int \cdots \int f(t) e^{it\cdot\xi}\, dt$$

$$= -i\xi_j \widehat{f}(\xi).$$

[Of course the "boundary terms" in the integration by parts vanish since $f \in C_c^\infty$.] The general case follows from a limiting argument (see Appendix I). $\square$

Proposition 2.1.3 *If $f \in L^1(\mathbb{R}^N)$ and $ix_j f \in L^1(\mathbb{R}^N)$, then*

$$(ix_j f)\widehat{\ } = \frac{\partial}{\partial \xi_j} \widehat{f}.$$

Proof. Differentiate under the integral sign.   □

Proposition 2.1.4 (The Riemann-Lebesgue Lemma)
If $f \in L^1(\mathbb{R}^N)$, then

$$\lim_{|\xi| \to \infty} |\widehat{f}(\xi)| = 0.$$

Proof. First assume that $g \in C_c^2(\mathbb{R}^N)$. We know that

$$\|\widehat{g}\|_{L^\infty} \le \|g\|_{L^1} \le C$$

and, for each j,

$$\left\| \xi_j^2 \widehat{g} \right\|_{L^\infty} = \left\| \left[\left(\frac{\partial^2}{\partial x_j^2} \right) g \right]\widehat{\ } \right\|_{L^\infty} \le \left\| \left(\frac{\partial^2}{\partial x_j^2} \right) g \right\|_{L^1} = C_j'.$$

Then $(1 + |\xi|^2)\widehat{g}$ is bounded. Therefore

$$|\widehat{g}(\xi)| \le \frac{C''}{1 + |\xi|^2} \overset{|\xi| \to \infty}{\longrightarrow} 0.$$

This proves the result for $g \in C_c^2$. [Notice that the same argument also shows that if $g \in C_c^{N+1}(\mathbb{R}^N)$ then $\widehat{g} \in L^1$.]

Now let $f \in L^1$ be arbitrary. By the results in Appendix I, there is a function $\psi \in C_c^2(\mathbb{R}^N)$ such that $\|f - \psi\|_{L^1} < \epsilon/2$.

Choose M so large that when $|\xi| > M$ then $|\widehat{\psi}(\xi)| < \epsilon/2$. Then, for $|\xi| > M$, we have

$$|\widehat{f}(\xi)| = |(f - \psi)\widehat{\ }(\xi) + \widehat{\psi}(\xi)| \le |(f - \psi)\widehat{\ }(\xi)| + |\widehat{\psi}(\xi)|$$

$$\le \|f - \psi\|_{L^1} + \frac{\epsilon}{2}$$

$$< \frac{\epsilon}{2} + \frac{\epsilon}{2} = \epsilon.$$

This proves the result.   □

Remark. The Riemann-Lebesgue lemma is intuitively clear when viewed in the following way. Fix an L^1 function f. An L^1 function is well-approximated by a continuous function (Appendix I), so we may as well suppose that f is continuous. But a continuous function is well-approximated by a smooth function, so we may as well suppose that f is smooth. On a small interval I—say of length $1/M$—a smooth function is nearly constant. So if we let $|\xi| \gg 2\pi M^2$, then the character $e^{i\xi \cdot x}$ will oscillate at least M times on I, and will therefore integrate against a constant to a value that is very nearly zero. As M becomes larger, this statement becomes more and more accurate. That is the Riemann-Lebesgue lemma.

Proposition 2.1.5 *Let* $f \in L^1(\mathbb{R}^N)$. *Then* $\widehat{f}$ *is uniformly continuous.*

Proof. Note that $\widehat{f}$ is continuous by the Lebesgue dominated convergence theorem:

$$\lim_{\xi \to \xi_0} \widehat{f}(\xi) = \lim_{\xi \to \xi_0} \int f(x)e^{ix \cdot \xi}\, dx = \int \lim_{\xi \to \xi_0} f(x)e^{ix \cdot \xi}\, dx = \widehat{f}(\xi_0).$$

Since $\widehat{f}$ also vanishes at ∞, the result is immediate. □

Let $C_0(\mathbb{R}^N)$ denote the continuous functions on $\mathbb{R}^N$ that vanish at ∞. Equip this space with the supremum norm. Then our results show that the Fourier transform maps L^1 to C_0 continuously, with operator norm 1 (Proposition 2.1.1). We shall show in Proposition 2.3.8 that it is *not* onto.

Later in this chapter (Section 2.3), we shall examine the action of the Fourier transform on L^2.

2.2 Invariance and Symmetry Properties of the Fourier Transform

The three Euclidean groups that act naturally on $\mathbb{R}^N$ are

- rotations
- dilations
- translations

Certainly a large part of the utility of the Fourier transform is that it has natural invariance properties under the actions of these three groups. We shall now explicitly describe those properties.

We begin with the orthogonal group $O(N)$; an $N \times N$ matrix is *orthogonal* if it has real entries and its rows form an orthonormal system of vectors. A *rotation* is an orthogonal matrix with determinant 1 (also called a *special orthogonal* matrix).

Proposition 2.2.1 *Let ρ be a rotation of $\mathbb{R}^N$. We define $\rho f(x) = f(\rho(x))$. Then we have the formula*

$$\widehat{\rho f} = \rho \widehat{f}.$$

Proof. Remembering that ρ is orthogonal and has determinant 1, we calculate that

$$
\begin{aligned}
\widehat{\rho f}(\xi) &= \int (\rho f)(t) e^{it \cdot \xi} \, dt = \int f(\rho(t)) e^{it \cdot \xi} \, dt \\
&\overset{(s = \rho(t))}{=} \int f(s) e^{i\rho^{-1}(s) \cdot \xi} \, ds = \int f(s) e^{is \cdot \rho(\xi)} \, ds \\
&= \widehat{f}(\rho \xi) = \rho \widehat{f}(\xi).
\end{aligned}
$$

Here we have used the fact that $\rho^{-1} = {}^t\rho$ for an orthogonal matrix. The proof is complete.   □

Definition 2.2.2 For $\delta > 0$ and $f \in L^1(\mathbb{R}^N)$ we set $\alpha_\delta f(x) = f(\delta x)$ and $\alpha^\delta f(x) = \delta^{-N} f(x/\delta)$. These are the dual *dilation operators* of Euclidean analysis.

Proposition 2.2.3 *The dilation operators interact with the Fourier transform as follows:*

$$(\alpha_\delta f)\widehat{\ } = \alpha^\delta \left(\widehat{f}\right)$$

$$\widehat{\alpha^\delta f} = \alpha_\delta \left(\widehat{f}\right).$$

Proof. We calculate that

$$
\begin{aligned}
(\alpha_\delta f)\widehat{\ }(\xi) &= \int (\alpha_\delta f)(t)e^{it\cdot\xi}\,dt \\
&= \int f(\delta t)e^{it\cdot\xi}\,dt \\
&\overset{(s=\delta t)}{=} \int f(s)e^{i(s/\delta)\cdot\xi}\delta^{-N}\,ds \\
&= \delta^{-N}\widehat{f}(\xi/\delta) \\
&= \left(\alpha^\delta(\widehat{f})\right)(\xi).
\end{aligned}
$$

That proves the first assertion. The proof of the second is similar.   □

For any function f on $\mathbb{R}^N$ and $a \in \mathbb{R}^N$ we define $\tau_a f(x) = f(x - a)$. Clearly τ_a is a *translation operator*.

Proposition 2.2.4 *If $f \in L^1(\mathbb{R}^N)$, then*

$$\widehat{\tau_a f}(\xi) = e^{ia\cdot\xi}\,\widehat{f}(\xi)$$

and

$$[\tau_a\{\widehat{f}\}](\xi) = \left[e^{-ia\cdot t}f(t)\right]\widehat{\ }(\xi).$$

Proof. For the first equality, we calculate that

$$
\begin{aligned}
\widehat{\tau_a f}(\xi) &= \int_{\mathbb{R}^N} e^{ix\cdot\xi}(\tau_a f)(x)\,dx \\
&= \int_{\mathbb{R}^N} e^{ix\cdot\xi} f(x - a)\,dx
\end{aligned}
$$

$$\overset{(x-a)=t}{=} \int_{\mathbb{R}^N} e^{i(t+a)\cdot\xi} f(t)\, dt$$

$$= e^{ia\cdot\xi} \int_{\mathbb{R}^N} e^{it\cdot\xi} f(t)\, dt$$

$$= e^{ia\cdot\xi} \widehat{f}(\xi).$$

The second identity is proved similarly. $\qquad\square$

Much of classical harmonic analysis—especially in this century—concentrates on translation-invariant operators. An operator T on functions is called *translation-invariant* if

$$T(\tau_a f)(x) = (\tau_a T f)(x)$$

for every x.[1] It is a basic fact that any translation-invariant integral operator is given by convolution with a kernel k (see also Lemma 1.6.8). We shall not prove this general fact (but see [STG1]), because the kernels that we need in the sequel will be explicitly calculated in context.

Proposition 2.2.5 *For $f \in L^1(\mathbb{R}^N)$ we let $\widetilde{f}(x) = f(-x)$. Then $\widehat{\widetilde{f}} = \widetilde{\widehat{f}}$.*

Proof. We calculate that

$$\widehat{\widetilde{f}}(\xi) = \int \widetilde{f}(t) e^{it\cdot\xi}\, dt = \int f(-t) e^{it\cdot\xi}\, dt$$

$$= \int f(t) e^{-it\cdot\xi}\, dt = \widehat{f}(-\xi) = \widetilde{\widehat{f}}(\xi). \qquad\square$$

Proposition 2.2.6 *We have*

$$\overline{\widehat{f}} = \widehat{\overline{\widetilde{f}}}.$$

---

[1] It is perhaps more accurate to say that such an operator *commutes with translations*. However, the terminology "translation-invariant" is standard.

Proof. We calculate that

$$\widehat{\overline{f}}(\xi) = \int \overline{f}(t)e^{it\cdot\xi}\,dt = \overline{\int f(t)e^{-it\cdot\xi}\,dt} = \overline{\widehat{f}(-\xi)} = \overline{\widetilde{\widehat{f}}}(\xi). \qquad \square$$

Proposition 2.2.7 *If* $f, g \in L^1$, *then*

$$\int \widehat{f}(\xi)g(\xi)\,d\xi = \int f(\xi)\widehat{g}(\xi)\,d\xi.$$

Proof. This is a straightforward change in the order of integration:

$$\int \widehat{f}(\xi)g(\xi)\,d\xi = \int \int f(t)e^{it\cdot\xi}\,dt\,g(\xi)\,d\xi$$

$$= \int \int g(\xi)e^{it\cdot\xi}\,d\xi\,f(t)\,dt$$

$$= \int \widehat{g}(t)f(t)\,dt.$$

Here we have applied the Fubini-Tonelli theorem. $\square$

We conclude this section with a brief discussion of homogeneity. Let $\beta \in \mathbb{R}$. We say that a function f on $\mathbb{R}^N$ (or, sometimes, on $\mathbb{R}^N \setminus \{0\}$) is *homogeneous of degree* β if, for each x and each $\lambda > 0$,

$$f(\lambda x) = \lambda^\beta f(x).$$

The typical example of a function homogeneous of degree β is $f(x) = |x|^\beta$, but this is not the only example. In fact, let ϕ be any function on the unit sphere of $\mathbb{R}^N$. Now set

$$f(x) = |x|^\beta \cdot \phi\left(\frac{x}{|x|}\right), \qquad x \neq 0.$$

Then f is homogeneous of degree β.

Let f be a function that is homogeneous of degree β on $\mathbb{R}^N$ (or on $\mathbb{R}^N \setminus \{0\}$). Then f will not be in any L^p class, so we may not consider its Fourier transform in the usual sense. Nonetheless, we may define the *weak Fourier transform* by the condition

$$\int_{\mathbb{R}^N} \widehat{f}(\xi) \cdot \phi(\xi)\, d\xi = \int_{\mathbb{R}^N} f(\xi) \cdot \widehat{\phi}(\xi)\, d\xi \qquad (2.2.8)$$

for any Schwartz function ϕ (see Appendix II). In other words, $\widehat{f}$ is defined by this equality. [The reader familiar with the theory of differential equations will note the analogy with the definition of "weak solution."] Of course such a definition is justified by Proposition 2.2.7. The well-definedness of $\widehat{f}$ according to this new definition follows from Fourier inversion (see Section 2.3). Then we have

Proposition 2.2.9 *Let f be a function that is homogeneous of degree β. Then $\widehat{f}$ is homogeneous of degree $-\beta - N$.*

Proof. Calculate $\alpha_\delta(\widehat{f})$ using (2.2.8) and a change of variables. $\square$

Remark. Technically speaking, (2.2.8) only defines $\widehat{f}$ as a distribution, or generalized function. Extra considerations are necessary to determine whether $\widehat{f}$ is a function.

2.3   Convolution and Fourier Inversion

Now we study how the Fourier transform respects the convolution of functions (see Section 0.2 for the definition).

Proposition 2.3.1 *If $f, g \in L^1$, then*

$$\widehat{f * g} = \widehat{f} \cdot \widehat{g}.$$

Proof. We calculate that

$$\widehat{f * g}(\xi) = \int (f * g)(t) e^{it \cdot \xi}\, dt = \int \int f(t - s) g(s)\, ds\, e^{it \cdot \xi}\, dt$$

$$= \int \int f(t - s) e^{i(t-s) \cdot \xi}\, dt\, g(s) e^{is \cdot \xi}\, ds = \widehat{f}(\xi) \cdot \widehat{g}(\xi).$$

The reader may note that the rather general form of Fubini's theorem, to which we alluded in Section 0.2, is needed to justify the change in the order of integration. $\square$

The Inverse Fourier Transform

Our goal is to be able to recover f from $\widehat{f}$. This program entails several technical difficulties. First, we need to know that the Fourier transform is one-to-one in order to have any hope of success. Secondly, we would like to say that

$$f(t) = c \cdot \int \widehat{f}(\xi) e^{-it \cdot \xi} \, d\xi. \tag{2.3.2}$$

But in general the Fourier transform $\widehat{f}$ of an L^1 function f is not integrable (just calculate the Fourier transform of $\chi_{[0,1]}$)—so the expression on the right of (2.3.1) does not necessarily make any sense. To handle this situation, we will construct a family of *summability kernels* G_ϵ having the following properties:

(2.3.3) $G_\epsilon * f \to f$ in the L^1 topology as $\epsilon \to 0$;

(2.3.4) $\widehat{G_\epsilon}(\xi) = e^{-\epsilon |\xi|^2 / 2}$;

(2.3.5) $G_\epsilon * f$ and $\widehat{G_\epsilon * f}$ are both integrable.

It will be useful to prove formulas about $G_\epsilon * f$ and then pass to the limit as $\epsilon \to 0^+$.

Lemma 2.3.6

$$\int_{\mathbb{R}^N} e^{-|x|^2} \, dx = (\sqrt{\pi})^N.$$

Proof. Breaking the integral into a product of 1-dimensional integrals, we see that it is enough to treat the case $N = 1$. Set $I = \int_{-\infty}^{\infty} e^{-t^2} \, dt$.

Then

$$I \cdot I = \int_{-\infty}^{\infty} e^{-s^2} ds \int_{-\infty}^{\infty} e^{-t^2} dt = \iint_{\mathbb{R}^2} e^{-|(s,t)|^2} ds dt$$

$$= \int_0^{2\pi} \int_0^{\infty} e^{-r^2} r \, dr d\theta = \pi.$$

Thus $I = \sqrt{\pi}$, as desired. $\qquad\square$

Remark. Although this is the most common method for evaluating $\int e^{-|x|^2} dx$, several other approaches are provided in [HEI].

Corollary 2.3.7

$$\int_{\mathbb{R}^N} \pi^{-N/2} e^{-|x|^2} \, dx = 1.$$

Now let us calculate the Fourier transform of $e^{-|x|^2}$. It is slightly more convenient to calculate the Fourier transform of $f(x) = e^{-|x|^2/2}$, and this we do.

It suffices to treat the 1-dimensional case because

$$\left(e^{-|x|^2/2}\right)\widehat{\ }(\xi) = \int_{\mathbb{R}^N} e^{-|x|^2/2} e^{ix\cdot\xi} dx$$

$$= \int_{\mathbb{R}} e^{-x_1^2/2} e^{ix_1\xi_1} dx_1 \cdots \int_{\mathbb{R}} e^{-x_N^2/2} e^{ix_N\xi_N} dx_N.$$

We thank J. Walker for providing the following argument (see also [FOL1, p. 242]):

By Proposition 2.1.3,

$$\frac{d\widehat{f}}{d\xi} = \int_{-\infty}^{\infty} ixe^{-x^2/2} e^{ix\xi} \, dx.$$

We now integrate by parts, with $dv = xe^{-x^2/2} dx$ and $u = ie^{ix\xi}$. The boundary terms vanish because of the presence of the rapidly decreasing exponential $e^{-x^2/2}$. The result is then

$$\frac{d\widehat{f}}{d\xi} = -\xi \int_{-\infty}^{\infty} e^{-x^2/2} e^{ix\xi} \, dx = -\xi \widehat{f}(\xi).$$

This is just a first-order linear ordinary differential equation for $\widehat{f}$. It is easily solved using the integrating factor $e^{\xi^2/2}$, and we find that

$$\widehat{f}(\xi) = \widehat{f}(0) \cdot e^{-\xi^2/2}.$$

But Lemma 2.3.6 (and a change of variable) tells us that $\widehat{f}(0) = \int f(x) \, dx = \sqrt{2\pi}$. In summary, on $\mathbb{R}^1$,

$$\widehat{e^{-x^2/2}}(\xi) = \sqrt{2\pi} e^{-\xi^2/2}; \tag{2.3.8}$$

in $\mathbb{R}^N$ we therefore have

$$(\widehat{e^{-|x|^2/2}})(\xi) = (\sqrt{2\pi})^N e^{-|\xi|^2/2}.$$

We can, if we wish, scale this formula to obtain

$$(\widehat{e^{-|x|^2}})(\xi) = \pi^{N/2} e^{-|\xi|^2/4}.$$

The function $G(x) = (2\pi)^{-N/2} e^{-|x|^2/2}$ is called the *Gauss-Weierstrass kernel*, or sometimes just the *Gaussian*. It is a summability kernel (see [KAT]) for the Fourier transform. Observe that

$$\widehat{G}(\xi) = e^{-|\xi|^2/2} = (2\pi)^{N/2} G(\xi). \tag{2.3.9}$$

On $\mathbb{R}^N$ we define

$$G_\epsilon(x) = \alpha^{\sqrt{\epsilon}}(G)(x) = \epsilon^{-N/2}(2\pi)^{-N/2} e^{-|x|^2/(2\epsilon)}.$$

Then

$$\widehat{G_\epsilon}(\xi) = \left(\alpha^{\sqrt{\epsilon}} G \right)^{\widehat{}}(\xi) = \alpha_{\sqrt{\epsilon}} \widehat{G}(\xi) = e^{-\epsilon|\xi|^2/2}$$

$$\widehat{\widehat{G_\epsilon}}(\xi) = \left(e^{-\epsilon|x|^2/2} \right)^{\widehat{}}(\xi)$$

$$= \left(\alpha_{\sqrt{\epsilon}} e^{-|x|^2/2} \right)^{\widehat{}} (\xi)$$

$$= \alpha^{\sqrt{\epsilon}} \left[(2\pi)^{N/2} e^{-|\xi|^2/2} \right]$$

$$= \epsilon^{-N/2} (2\pi)^{N/2} e^{-|\xi|^2/(2\epsilon)}$$

$$= (2\pi)^N G_\epsilon(\xi).$$

Observe that $\widehat{\widehat{G_\epsilon}}$ is, except for the factor of $(2\pi)^N$, the same as G_ϵ. This fact anticipates the Fourier inversion formula that we are gearing up to prove.

Now assume that $f \in C_c^{N+1}$. This implies in particular that f, $\widehat{f}$ are in L^1 and continuous (see the proof of the Riemann-Lebesgue lemma, Lemma 2.1.4). We apply Proposition 2.2.7 with $g = \widehat{G_\epsilon} \in L^1$ to obtain

$$\int f(x) \widehat{\widehat{G_\epsilon}}(x) \, dx = \int \widehat{f}(\xi) \widehat{G_\epsilon}(\xi) \, d\xi.$$

In other words,

$$\int f(x) (2\pi)^N G_\epsilon(x) \, dx = \int \widehat{f}(\xi) e^{-\epsilon|\xi|^2/2} \, d\xi. \tag{2.3.10}$$

Now $e^{-\epsilon|\xi|^2/2} \to 1$ as $\epsilon \to 0^+$, uniformly on compact sets. Thus $\int \widehat{f}(\xi) e^{-\epsilon|\xi|^2/2} \, d\xi \to \int \widehat{f}(\xi) \, d\xi$. That concludes our analysis of the right-hand side of (2.3.10).

Next observe that, for any $\epsilon > 0$,

$$\int G_\epsilon(x) \, dx = \int G_\epsilon(x) e^{ix \cdot 0} \, dx = \widehat{G_\epsilon}(0) = 1.$$

Thus the left side of (2.3.10) equals

$$(2\pi)^N \int f(x) G_\epsilon(x) \, dx = (2\pi)^N \int f(0) G_\epsilon(x) \, dx$$

$$+ (2\pi)^N \int [f(x) - f(0)] G_\epsilon(x) \, dx$$

$$\equiv A_\epsilon + B_\epsilon.$$

Now it is crucial to observe that the function G_ϵ has total mass 1, and that mass is more and more concentrated near the origin as $\epsilon \to 0^+$. Refer to Figure 1. As a result, $A_\epsilon \to (2\pi)^N f(0)$ and $B_\epsilon \to 0$. Altogether then,

$$\int f(x)(2\pi)^N G_\epsilon(x)\, dx \to (2\pi)^N f(0)$$

as $\epsilon \to 0^+$. Thus we have evaluated the limits of the left- and right-hand sides of (2.3.10). [Note that the argument just given is analogous to some of the proofs for the summation of Fourier series that were presented in Chapter 1.] We have proved the following:

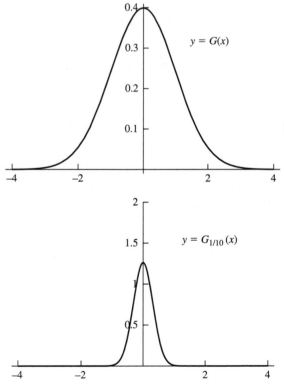

Figure 1.   Portrait of G_ϵ.

Proposition 2.3.11 (Gauss-Weierstrass Summation) *Suppose that* $f \in C_c^{N+1}(\mathbb{R}^N)$ *(this hypothesis is included to guarantee that $\widehat{f} \in L^1$).* *Then*

$$f(0) = \lim_{\epsilon \to 0^+} (2\pi)^{-N} \int \widehat{f}(\xi) e^{-\epsilon |\xi|^2/2} \, d\xi.$$

The method of Gauss-Weierstrass summation will prove to be crucial in some of our later calculations (see especially Chapter 3). However, in practice, it is convenient to have a result with a simpler and less technical statement. If f, $\widehat{f}$ are both known to be in L^1 (this is true, for

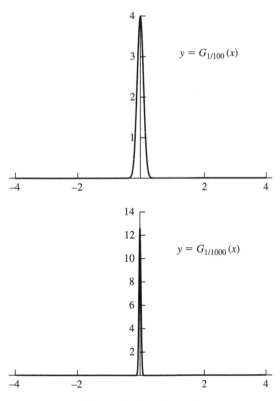

Figure 1. *(continued)*

example, if f has $(N + 1)$ derivatives in L^1), then a limiting argument gives the following standard result:

Theorem 2.3.12 *If f, $\widehat{f} \in L^1$ (and both are continuous), then*

$$f(0) = (2\pi)^{-N} \int \widehat{f}(\xi) \, d\xi. \qquad (2.3.12.1)$$

Figure 2 shows how $\widehat{G_\epsilon}$ flattens out as $\epsilon \to 0$, uniformly on compact sets, as $\epsilon \to 0^+$.

Of course there is nothing special about the point $0 \in \mathbb{R}^N$. We now exploit the compatibility of the Fourier transform with translations

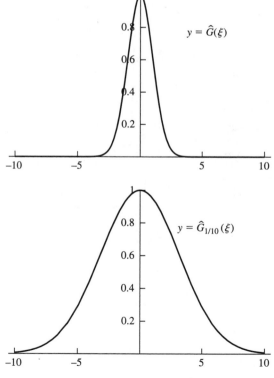

Figure 2. Portrait of $\widehat{G_\epsilon}$.

to obtain a more general formula. We apply formula (2.3.12.1) in our theorem to $\tau_{-y} f$: The result is

$$\left(\tau_{-y} f\right)(0) = (2\pi)^{-N} \int \widehat{(\tau_{-y} f)}(\xi)\, d\xi$$

or

Theorem 2.3.13 (The Fourier Inversion Formula) *If $f, \widehat{f} \in L^1$ (and if both $f, \widehat{f}$ are continuous), then for any $y \in \mathbb{R}^N$ we have*

$$f(y) = (2\pi)^{-N} \int \widehat{f}(\xi) e^{-iy\cdot\xi}\, d\xi.$$

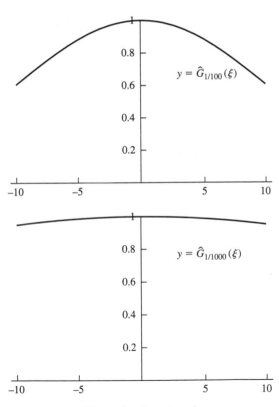

Figure 2. (*continued*)

Observe that this theorem tells us that if f, $\widehat{f}$ are both L^1, then f (being the inverse Fourier transform of an L^1 function) can be corrected on a set of measure zero to be continuous.

Corollary 2.3.14 *The Fourier transform is one-to-one. That is, if $f, g \in L^1$ and $\widehat{f} \equiv \widehat{g}$, then $f = g$ almost everywhere.*

Proof. Observe that $f - g \in L^1$ and $\widehat{f} - \widehat{g} \equiv 0 \in L^1$. The theorem allows us to conclude that $(f - g) * G_\epsilon = 0$ for all $\epsilon > 0$, and passage to the limit now yields the result. □

Even though we do not know the Fourier transform to be a surjective operator (onto the continuous functions vanishing at infinity), it is convenient to be able to make explicit reference to the inverse operation. Thus we *define*

$$\check{g}(x) = (2\pi)^{-N} \int g(\xi) e^{-ix \cdot \xi} \, d\xi$$

whenever $g \in L^1(\mathbb{R}^N)$. We call the operation $\check{}$ the *inverse Fourier transform*. Notice that the operation expressed in the displayed equation makes sense for any $g \in L^1$, regardless of the operation's possible role as an inverse to the Fourier transform. It is convenient to give the operation the italicized name.

In particular, observe that if the hypotheses of the Fourier inversion theorem are in force, then we see that

$$f(x) = \check{\widehat{f}}(x).$$

Since the Fourier transform is one-to-one, it is natural to ask whether it is onto. We have

Proposition 2.3.15 *The operator*

$$\widehat{} : L^1 \to C_0$$

is not onto.

Proof. For simplicity, we restrict attention to $\mathbb{R}^1$. Seeking a contradiction, we suppose that the operator is in fact surjective. Then the open mapping principle guarantees that there is a constant $C > 0$ such that

$$\|f\|_{L^1} \le C \cdot \|\widehat{f}\|_{\sup} \qquad \text{for all } f \in L^1. \tag{2.3.15.1}$$

On $\mathbb{R}^1$, let g be the characteristic function of the interval $[-1, 1]$. A calculation shows that the inverse Fourier transform of g is

$$\check{g}(t) = \frac{\sin t}{\pi t}.$$

We would like to say that $f = \check{g}$ violates (2.3.15.1). But this f is not in L^1, so such logic is flawed.

Instead we consider $h_\epsilon \equiv G_\epsilon * g$. Then h_ϵ, being the convolution of two L^1 functions, is certainly L^1. Moreover,

$$|h_\epsilon(x)| = \epsilon^{-1/2}(2\pi)^{-1/2} \int_{-1}^{1} e^{-|x-t|^2/2\epsilon} \, dt$$

$$= \epsilon^{-1/2}(2\pi)^{-1/2} e^{-x^2/2\epsilon} \int_{-1}^{1} e^{xt/\epsilon} e^{-t^2/2\epsilon} \, dt$$

$$\le C_\epsilon e^{-x^2/4\epsilon}.$$

So $h_\epsilon \in L^1 \cap C_0$.

In addition,

$$\check{h}_\epsilon(t) = 2\pi \cdot \check{G}_\epsilon(t) \cdot \check{g}(t) = e^{-\epsilon|t|^2/2} \cdot \check{g}(t) \to \check{g}(t)$$

pointwise. This convergence cannot be boundedly in L^1, otherwise Fatou's lemma would show that $\check{g} \in L^1$; and we know that that is false.

A contradiction now arises because

$$\|2\pi \cdot \check{G}_\epsilon \cdot \check{g}\|_{L^1} = \|(G_\epsilon * g)^{\check{}}\|_{L^1} \le C \cdot \|G_\epsilon * g\|_{L^\infty}$$

$$\le C \cdot \|G_\epsilon\|_{L^1} \cdot \|g\|_{L^\infty} = C.$$

As noted, the left side blows up as $\epsilon \to 0^+$. $\qquad\square$

Exercise. Imitate the proof just presented to show that the mapping

$$\widehat{}: L^1(\mathbb{T}) \to c_0,$$

which assigns to each integrable function on the circle its sequence of Fourier coefficients, is not onto.

The Fourier inversion formula is troublesome. First, it represents the function f as the superposition of *uncountably many* basis elements $e^{-ix\cdot\xi}$, none of which is in any L^p class. In particular, the Fourier transform does not localize well. The theory of wavelets (Chapter 7) is designed to address some of these shortcomings.

Plancherel's Formula

We refer the reader to Section 1.6 for our initial remarks about the quadratic Fourier theory. Now we give a more detailed treatment.

Proposition 2.3.16 (Plancherel) *If $f \in C_c^\infty(\mathbb{R}^N)$, then*

$$(2\pi)^{-N} \int |\widehat{f}(\xi)|^2 \, d\xi = \int |f(x)|^2 \, dx.$$

Proof. Define $g = f * \widetilde{\overline{f}} \in C_c^\infty(\mathbb{R}^N)$. Then

$$\widehat{g} = \widehat{f} \cdot \widehat{\widetilde{\overline{f}}} = \widehat{f} \cdot \widetilde{\widehat{\overline{f}}} = \widehat{f} \cdot \widetilde{\overline{\widehat{f}}} = \widehat{f} \cdot \overline{\widetilde{\widehat{f}}} = \widehat{f} \cdot \overline{\widehat{f}} = |\widehat{f}|^2. \quad (2.3.16.1)$$

Now

$$g(0) = f * \widetilde{\overline{f}}\,(0) = \int f(-t)\overline{f}(-t) \, dt$$

$$= \int f(t)\overline{f}(t) \, dt = \int |f(t)|^2 \, dt.$$

By Fourier inversion and formula (2.3.16.1) we may now conclude that

$$\int |f(t)|^2 \, dt = g(0) = (2\pi)^{-N} \int \widehat{g}(\xi) \, d\xi = (2\pi)^{-N} \int |\widehat{f}(\xi)|^2 \, d\xi.$$

That is the desired formula. $\qquad\square$

Definition 2.3.17 For any $f \in L^2(\mathbb{R}^N)$, the Fourier transform of f can be defined in the following fashion: Let $f_j \in C_c^\infty$ satisfy $f_j \to f$ in the L^2 topology (see Appendix I). It follows from the proposition that $\{\widehat{f_j}\}$ is Cauchy in L^2. Let g be the L^2 limit of this latter sequence. We set $\widehat{f} = g$.

It is easy to check that this definition of $\widehat{f}$ is independent of the choice of sequence $f_j \in C_c^\infty$ and that

$$(2\pi)^{-N} \int |\widehat{f}(\xi)|^2 \, d\xi = \int |f(x)|^2 \, dx.$$

We next record the "polarized form" of Plancherel's formula:

Proposition 2.3.18 If $f, g \in L^2(\mathbb{R}^N)$, then

$$\int f(t) \cdot \overline{g(t)} \, dt = (2\pi)^{-N} \int \widehat{f}(\xi)\overline{\widehat{g}}(\xi) \, d\xi.$$

Proof. The proof consists of standard tricks from algebra: First assume that f, g are real-valued. Apply the Plancherel formula that we proved to the function $(f + g)$ and then again to the function $(f - g)$ and then subtract.

The case of complex-valued f and g is treated by applying the Plancherel formula to $f + g$, $f - g$, $f + ig$, $f - ig$ and combining.

$\square$

Exercises. Restrict attention to dimension 1. Consider the Fourier transform $\mathcal{F}$ as a bounded linear operator on the Hilbert space $L^2(\mathbb{R})$. Prove that the four roots of unity (suitably scaled) are eigenvalues of $\mathcal{F}$. [*Hint:* What happens when you compose the Fourier transform with itself four times?]

Which functions in L^2 are invariant (up to a scalar factor) under the Fourier transform? We know that $(ixf)\widehat{}(\xi) = (\widehat{f})'(\xi)$ and $(f')\widehat{}(\xi) = -i\xi \widehat{f}(\xi)$. As a result, the differential operator $d^2/dx^2 - x^2 I$

is invariant under the Fourier transform. It seems plausible that any solution of the differential equation

$$\frac{d^2}{dx^2}\phi - x^2\phi = \lambda\phi,\qquad(2.3.19)$$

for λ a suitable constant, will also be mapped by the Fourier transform to itself. Since the function $e^{-x^2/2}$ is (up to a constant) mapped to itself by the Fourier transform, it is natural to guess that equation (2.3.19) would have solutions involving this function. Thus perform the change of notation $\phi(x) = e^{-x^2/2} \cdot \Phi(x)$. Guess that Φ is a polynomial, and derive recursions for the coefficients of that polynomial. These polynomials are called the *Hermite polynomials*. A full treatment of these matters appears in [WIE, pp. 51–55] or in [FOL1, p. 248].

You may also verify that the polynomials Φ that you find form (after suitable normalization) an orthonormal basis for L^2 when calculated with respect to the measure $d\mu = \sqrt{2}e^{-x^2/2}\,dx$. For details, see [WIE].

We now know that the Fourier transform $\mathcal{F}$ has the following mappings properties:

$$\mathcal{F}: L^1 \to L^\infty$$
$$\mathcal{F}: L^2 \to L^2.$$

These are both bounded operations. It follows that $\mathcal{F}$ is defined on L^p for $1 < p < 2$ (since, for such p, $L^p \subseteq L^1 + L^2$).

The Riesz-Thorin interpolation theorem (Appendix III) now allows us to conclude that

$$\mathcal{F}: L^p \to L^{p'}, \quad 1 \le p \le 2,$$

where $p' = p/(p - 1)$. If $p > 2$, then $\mathcal{F}$ does not map L^p into any nice function space. The precise norm of $\mathcal{F}$ on L^p, $1 \le p \le 2$, has been computed by Beckner [BEC].

2.4   The Uncertainty Principle

In this section we describe and prove a fundamental result of Fourier analysis that is known as the uncertainty principle. In fact this theorem was "discovered" by W. Heisenberg in the context of quantum mechanics. Expressed colloquially, the uncertainty principle says that it is not possible to know both the position and the momentum of a particle at the same time. Expressed more precisely, the uncertainty principle says that the position and the momentum cannot be simultaneously localized. A charming discussion of the physical ideas appears in [STR1] and in [FOS].

For us, the uncertainty principle is an incisive fact about the Fourier transform. We shall formulate both a qualitative version and a quantitative version. Recall that a function f is said to be of *compact support* if there is a compact set K such that $f(x) = 0$ when $x \notin K$.

Theorem 2.4.1 (Uncertainty Principle; Qualitative Version) *Let f be a continuous function on $\mathbb{R}^N$ and assume that f has compact support. If $\widehat{f}$ also has compact support, then $f \equiv 0$.*

Proof. For simplicity, we restrict attention to dimension $N = 1$. Consider the function of a complex variable given by

$$\Phi : z \mapsto \int_{\mathbb{R}} f(t) e^{it \cdot z} \, dt.$$

An application of Morera's theorem (see [GRK]), or differentiation under the integral sign, shows that Φ is an entire function—that is, a holomorphic function on the entire complex plane. [Note that this is where we use the fact that f is compactly supported; were it not, there would be trouble with convergence of the integral.]

Write $z = x + iy$ and restrict Φ to the set where $y = 0$. Then

$$\Phi(x + i0) = \widehat{f}(x).$$

But our hypothesis is that $\widehat{f}$ *vanishes* on an entire ray. Thus we have an entire holomorphic function Φ that vanishes on a ray; in particular, it vanishes on a set with an interior accumulation point. Thus $\Phi \equiv 0$. By Fourier uniqueness, $f \equiv 0$. $\qquad\square$

The uncertainty principle has profound consequences for the theory of partial differential equations. As C. Fefferman explains in [FEF3], one precise quantitative formulation of the principle says that if a function f, satisfying $\int |f|^2 = 1$, is "mostly concentrated" in an interval of length $2\delta_x$, and if its Fourier transform is "mostly concentrated" in an interval of length $2\delta_\xi$, then it must be that $\delta_x \cdot \delta_\xi \geq 1$. Here is what this means in practice:

The sort of analysis that one usually performs when analyzing an elliptic partial differential operator is to endeavor to break up (x, ξ)-space into boxes (here x is the spatial variable that describes the domain of f, and ξ is the "phase variable" that describes the domain of $\widehat{f}$). The uncertainty principle puts important restrictions on the shapes of those boxes. This information in turn determines the nature of the Gårding inequality, which is the fundamental fact about elliptic partial differential operators. Again, see [FEF3] for the complete picture.

We now present one version of the quantitative uncertainty principle. When $\|f\|_{L^2} = 1$, it is convenient for us to use the notation

$$\overline{x} = \int_{\mathbb{R}^N} x |f(x)|^2 \, dx$$

and

$$\text{Var } f \equiv \inf_{y \in \mathbb{R}^N} \int_{\mathbb{R}^N} |x - y|^2 |f(x)|^2 \, dx$$

when these expressions make sense. [It can be checked that, when $\overline{x}$ exists, then the infimum that defines Var f is assumed when $y = \overline{x}$.]

Theorem 2.4.2 (Uncertainty Principle; Quantitative Version) *Let f be an L^2 function on $\mathbb{R}^N$ normalized so that $\int |f|^2 = 1$. Then*

$$[\text{Var } f] \cdot [\text{Var } \widehat{f}] \geq \frac{N^2}{4}.$$

Observe that the variance Var f measures the least extent to which f is concentrated about any point—where the notion of mean is determined with respect to the weight $|f|^2$.

Proof of the Theorem. For simplicity, we restrict attention to dimension 1. We also restrict attention to smooth functions f such that $f, xf, df/dx$ lie in L^2.

Following [STR1], we let A denote the operator $i(d/dx)$ and we let B denote the operator of multiplication by x. We suppose for simplicity that $\overline{A} \equiv \langle Af, f \rangle = 0$ and $\overline{B} = \langle Bf, f \rangle = 0$. In particular, $\overline{x} = 0$.

We use the symbol $[A, B]$ to denote the so-called *commutator* of A and B: $[A, B] = AB - BA$. Then

$$[A, B]\phi(x) = [AB\phi - BA\phi](x)$$
$$= i\frac{d}{dx}[x\phi(x)] - xi\frac{d}{dx}\phi(x)$$
$$= i\phi(x).$$

We may rewrite this result as $-i[A, B] = I$, where I denotes the identity operator.

Recall that the inner product on L^2 is

$$\langle f, g \rangle = \int_{\mathbb{R}^N} f(x) \cdot \overline{g(x)}\, dx.$$

You may calculate as an exercise that, if g is also a smooth function such that $g, xg, dg/dx \in L^2$, then

$$\langle Af, g \rangle = \langle f, Ag \rangle$$

(just integrate by parts, or see [FOS] for a hint) and

$$\langle Bf, g \rangle = \langle f, Bg \rangle.$$

Now we have, for a function f with $\int |f|^2 = 1$ (as well as $xf, df/dx \in L^2$), that

$$1 = \int_{\mathbb{R}} |f(x)|^2 dx$$
$$= \langle f, f \rangle$$
$$= \langle -i(AB - BA)f, f \rangle$$
$$= -i\langle Bf, Af \rangle + i\langle Af, Bf \rangle$$
$$= 2\mathrm{Re}\left(i\langle Af, Bf \rangle \right).$$

Now we apply the Cauchy-Schwarz inequality (or Hölder's inequality when $p = q = 2$). The result is

$$1 = 2|\mathrm{Re}(i\langle Af, Bf \rangle)|$$
$$\leq 2|\langle Af, Bf \rangle|$$
$$\leq 2\langle Af, Af \rangle^{1/2}\langle Bf, Bf \rangle^{1/2}$$

hence

$$\frac{1}{4} \leq \langle Af, Af \rangle\langle Bf, Bf \rangle.$$

Observe that the expression $\langle Bf, Bf \rangle$ is nothing other than Var f (since $\overline{x} = 0$). Using Plancherel's theorem together with Propositions 2.1.2 and 2.1.3, we see also that $\langle Af, Af \rangle$ equals Var $\widehat{f}$. Thus we have obtained the result.   □

The uncertainty principle is an essential feature of virtually any version of Fourier analysis. Wavelet theory (Chapter 7) endeavors to give up certain features of classical Fourier analysis in order to minimize the effect of the uncertainty principle—so that localization both in the space variable and the phase variable is possible. In particular, wavelet theory gives us a basis whose elements have an uncertainty product (as in Theorem 2.4.2) that is reasonably close to being optimal—in the sense that one has best possible decay of both f and $\widehat{f}$. For more details, see formula (3.9) in [FOS] and the discussion pertaining thereto.

We shall explore some of these ideas at the end of the book.

CHAPTER 3

Multiple Fourier Series

3.1 Various Methods of Partial Summation

In Chapter 1 we learned that a suitable method for summing the Fourier series of a function f on the circle is to define partial sums

$$S_N f(x) \equiv \sum_{j=-N}^{N} \widehat{f}(j) e^{ijx}.$$

Recall that questions of convergence of $S_N f$ to f reduced to the study of the Hilbert transform; the Hilbert transform, in turn, corresponds (roughly speaking) to the summation operation

$$f \longmapsto \sum_{j=0}^{\infty} \widehat{f}(j) e^{ijx}.$$

Running this reasoning in reverse, it is possible to prove that almost *any* reasonable definition of partial sum will give a tractable theory for Fourier analysis on the circle group. More precisely, if $\alpha(N), \beta(N)$ are both positive, strictly increasing functions that take integer values and if we define

$$\widetilde{S}_N f(x) \equiv \sum_{j=-\alpha(N)}^{\beta(N)} \widehat{f}(j) e^{ijx}$$

then any theorem that we proved in Chapter 1 for the classical partial summation operators S_N (or for any summation operators formed from the S_N) will also hold for the $\widetilde{S}_N$. We leave the details of this assertion as an exercise for the interested reader.

Matters are quite different for Fourier series in several variables. Any intuition that multivariable real analysis is just like one-variable analysis—but with more notation and bookkeeping—is quickly dispelled when one begins to consider different methods of summation. In the next section, we give some numerical examples that begin to suggest how truly complex matters can be. Although the theory in 3 dimensions and higher is distinctly more complicated than that in 2 dimensions, we can already give adequate illustration of the problems that arise by confining our attention to dimension 2. This we do.

Let us define several notions of partial summation. All of these are, in a sense, inspired by the classical partial-summation operators S_N from dimension 1. But each takes into account the symmetries of space in a different manner. And each has associated to it a distinct analysis.

Our venue now is $\mathbb{T}^2 \equiv \mathbb{T} \times \mathbb{T}$. We perform integration on $\mathbb{T}^2$ by modeling it (in analogy with what we did in the one-variable theory) with $[0, 2\pi) \times [0, 2\pi)$. Thus $f \in L^1(\mathbb{T}^2)$ precisely when $f(e^{is}, e^{it})$ lies in $L^1([0, 2\pi) \times [0, 2\pi))$. For such an f, we define

$$\widehat{f}(j, k) \equiv \frac{1}{(2\pi)^2} \int_0^{2\pi} \int_0^{2\pi} f(s, t) e^{-i(sj+tk)} \, ds \, dt$$

$$= \frac{1}{(2\pi)^2} \int_0^{2\pi} \int_0^{2\pi} f(s, t) e^{-i(s,t)\cdot(j,k)} \, ds \, dt.$$

Here $-\infty < j, k < \infty$ are integers. The issue is how to sum the terms $\widehat{f}(j, k) e^{i(jx+ky)}$ in order to recover f. In what follows, we use $|(x, y)|$ to denote $\sqrt{x^2 + y^2}$ and $|(j, k)|$ to denote $\sqrt{j^2 + k^2}$.

The first method of partial summation that we shall introduce is commonly known as *spherical summation* (although "ball summation" would be more accurate in terms of standard mathematical usage). For this method we define

$$S_R^{\text{sph}} f(x, y) \equiv \sum_{|(j,k)| \leq R} \widehat{f}(j, k) e^{i(jx+ky)}.$$

The provenance of the name is now obvious: We let the partial sum be the aggregate of those terms with index lying in the ball $\overline{B}(0, R)$ in index space. The question that we would like to pose about S_R^{sph} is whether

$$S_R^{\text{sph}} f(x, y) \rightarrow f(x, y) \quad \text{as } R \rightarrow \infty$$

either pointwise, or in norm, or in the sense of some summability operation. Until the time of a startling counterexample of Fefferman [FEF2], which we shall describe later in this chapter, spherical summation was thought to be the most natural method for summing multiple Fourier series. It is also among the most difficult summation methods to study.

The second method of partial summation that we shall study is "square summability." Now we define

$$S_M^{\text{sq}} f(x, y) \equiv \sum_{\substack{|j| \leq M \\ |k| \leq M}} \widehat{f}(j, k) e^{i(jx+ky)}.$$

Notice that in this summation method we let a partial sum be the aggregate of those terms with indices lying in a *closed square* of side $2M$ and center the origin in index space. Again, the basic question will be whether

$$S_M^{\text{sq}} f(x, y) \rightarrow f(x, y) \quad \text{as } M \rightarrow \infty.$$

Spherical and square summability are perhaps the two most natural summation methods for 2-dimensional Fourier series. But there are others that play a useful role in analysis. We shall mention three of these.

First we define *restricted rectangular convergence*. Fix a number $E > 1$, where "E" denotes "eccentricity." We consider partial sums of the form

$$S_{(m,n)}^{\text{rect}} f(x, y) \equiv \sum_{\substack{|j| \leq m \\ |k| \leq n}} \widehat{f}(j, k) e^{i(jx + ky)}$$

for integers $m, n \geq 0$. For restricted rectangular convergence, we mandate that $1/E \leq m/n \leq E$. In other words, we require that the rectangle of indices over which we are summing not be too eccentric (i.e., too long and narrow). Then the restricted rectangular convergence question is whether

$$\lim_{N \to \infty} \sup_{\substack{m, n > N \\ 1/E < m/n < E}} |S_{(m,n)}^{\text{rect}} f(x, y) - f(x, y)| = 0$$

(either pointwise or in norm). Notice that, in studying restricted rectangular convergence, we require that the rectangles of indices over which we are summing not deviate too far from a square (the bound on the deviation is given by E).

For unrestricted rectangular convergence, we ask whether

$$\lim_{N \to \infty} \sup_{m, n > N} |S_{(m,n)}^{\text{rect}} f(x, y) - f(x, y)| = 0.$$

Observe that here we consider all rectangles (with sides parallel to the axes) of indices in our summation process. There is no restriction on their eccentricities.

Finally, we consider polygonal summation. Fix a closed, convex polygon P (with only finitely many sides) in the plane (more precisely, in index space) that contains the origin. See Figure 1. For $R > 0$ we let $R \cdot P = \{(Rx, Ry) : (x, y) \in P\}$. We define the *polygonal partial summation operator with index R* to be

$$S_R^{\text{poly}, P} f(x, y) = \sum_{(j,k) \in R \cdot P} \widehat{f}(j, k) e^{i(jx + ky)}.$$

The basic question is whether

$$\lim_{R \to \infty} S_R^{\text{poly}, P} f(x, y) = f(x, y)$$

either pointwise, or in norm, or by some summation method.

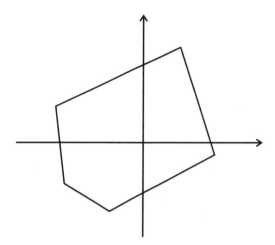

Figure 1.   Polygonal summation.

Of course square partial summation is a special case of polygonal partial summation. Once we develop the appropriate machinery for studying these partial-summation operators, we shall see that square summation and polygonal summation are quite similar—in fact they may all be studied in essentially the same way. But unrestricted rectangular summation and spherical summation stand apart. [Restricted rectangular convergence is somewhere between these extremes.] Unrestricted summation is different because of the lack of control on eccentricities; spherical summation stands apart because the tangent lines to the boundary of a sphere (in 2 dimensions, a circle) point in too many different directions. Notice, by contrast, that the tangent lines to the boundary of a square, or of a polygon, can point in just finitely many different directions. The functional analysis machinery that we shall develop below to study these problems will make these observations crystal clear.

In what follows, we will use the phrase "coefficient region" to mean the geometric region (usually in the plane) that contains the Fourier coefficient indices over which we are summing. For spherical summation the coefficient regions will be closed discs, for rectangular summation, they will be closed rectangles, and so forth. Context will make clear in each instance precisely what is meant.

3.2 Examples of Different Types of Summation

Now we give some examples of double numerical series that illustrate how the different partial-summation methods can lead to different results. These are drawn from [ASH1]. In each example we will consider a series

$$\sum_{j,k} a_{j,k},$$

where the a_{jk} are numbers. The various summation methods discussed above will be applied to this numerical series. For instance,

$$S_R^{\text{sph}} \equiv \sum_{j^2+k^2 \leq R^2} a_{j,k}.$$

Example 3.2.1 For $k > 0$ let

(3.2.1.1) $a_{5k,0} = -k$;

(3.2.1.2) $a_{4k,3k} = k$;

(3.2.1.3) all other $a_{\ell,m}$ equal 0.

Then, for any integer $k \geq 1$, the closed square (in index space) with center $(0, 0)$ and side $2 \cdot (5k - 1)$ contains the indices

$$a_{5,0}, a_{10,0}, \ldots, a_{5(k-1),0}$$

and

$$a_{4,3}, a_{8,6}, \ldots, a_{4k,3k}.$$

In particular, the square contains at least one more index of the second type than of the first type. Hence $S_{5k-1}^{\text{sq}} \geq k$, and the square partial sums diverge to ∞.

However, note that any closed ball (in index space) centered at the origin that contains $(5j, 0)$ also contains $(4j, 3j)$ and conversely. As a result, $S_R^{\text{sph}} = 0$ for every R and the series *is* spherically convergent.

Since the series is square divergent, it is also restrictedly rectangularly divergent and certainly unrestrictedly rectangularly divergent. However it can be polygonally convergent for some polygons but not for others. To see this last assertion, let P be the closed regular octagon centered at the origin, with four of its sides parallel to the axes (Figure 1). Then any dilate $R \cdot P$ that contains a point $(5j, 0)$ will also contain $(4j, 3j)$, and conversely. As a result, all the polygonal partial sums—for this particular polygon—will be 0. Thus the series is polygonally convergent when P is the regular octagon. We have already produced a polygon (the square) for which divergence obtains. The interested reader may produce other examples of polygons whose

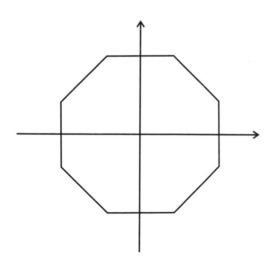

Figure 1. Octagonal summation.

summation schemes will lead to divergence or convergence of the series. $\qquad\square$

Example 3.2.2 For $k > 0$, let

(3.2.2.1) $a_{k,k} = -k$;

(3.2.2.2) $a_{k,0} = k$;

(3.2.2.3) all other $a_{\ell,m}$ equal 0.

Then it is plain (see the reasoning in the last example) that the series is convergent, by the method of square summation, to 0. However, by contrast,

$$S^{\text{rect}}_{(k,k-1)} = k$$

for every k. Thus the series is not convergent by the method of restricted rectangular summation. It is also not spherically convergent since (using $[\alpha]$ to denote the greatest integer in α)

$$S^{\text{sph}}_k = \sum_{j=1}^{k} j - \sum_{j=1}^{[k/\sqrt{2}]} j$$

$$= \frac{k(k+1)}{2} - \frac{[k/\sqrt{2}]([k/\sqrt{2}]+1)}{2}$$

$$\geq \frac{k^2}{8} \to \infty$$

as $k \to \infty$. $\qquad\square$

Example 3.2.3 For $k > 0$, let

(3.2.3.1) $a_{k^2,0} = k$;

(3.2.3.2) $a_{k^2,k} = -k$;

(3.2.3.3) all other $a_{\ell,m}$ equal 0.

The series then converges restrictedly rectangularly to 0, since if we restrict attention to rectangles of eccentricity at most E then, when $k \gg E$ it must be that any rectangle that contains $(k^2, 0)$ (in index space) must also contain (k^2, k). But notice that

$$S^{\text{rect}}_{(k^2, k-1)} = k$$

(of course the rectangles being considered here do *not* have bounded eccentricity). Hence the series diverges unrestrictedly rectangularly. The series also diverges spherically since

$$S^{\text{sph}}_{k^2} = \sum_{j=1}^{k} j - \sum_{j=1}^{k-1} j = k \to \infty$$

as $k \to \infty$.   □

Example 3.2.4 For $k > 0$, let

(3.2.4.1) $a_{k,0} = k$;

(3.2.4.2) $a_{k,1} = -k$;

(3.2.4.3) all other $a_{\ell,m}$ equal 0.

Then the series converges unrestrictedly rectangularly. That is, any sufficiently large rectangle (both dimensions must be large) will contain an equal number of k and $-k$ terms, so the partial sums will be 0. But the terms are plainly unbounded, and the spherical partial sums are unbounded. So the series diverges spherically.   □

These examples teach us that, apart from the trivial implications

unrestricted rectangular convergence $\Rightarrow$

restricted rectangular convergence $\Rightarrow$

square convergence,

all other implications among the different convergence methods are false. *Notice that the sentence just enunciated has only been shown to be true for numerical series, and not for series which we know arise as Fourier series.* It is conceivable that, when we are dealing with "almost everywhere" convergence results, or norm convergence results for multiple Fourier series, then other implications may hold; however, the rule of thumb of the experts [ASH1] is that "double trigonometric series [are] usually just as bad as double numerical series."

We shall learn in the next section that Fourier multipliers may be used as a unifying theme for the study of the convergence of multiple Fourier series (in fact this was true for Fourier series of one variable as well, but the role of the multiplier was more tacit in our Chapter 1).

3.3 Fourier Multipliers and the Summation of Series

We shall learn in this section—by way of a suitable Poisson summation formula plus certain limiting arguments—that the question of whether spherical summation works in the L^p norm for a given p is completely equivalent to whether the operator

$$\mathcal{M}_B : f \longmapsto \left[\chi_{B(0,1)} \cdot \widehat{f} \right]^{\vee}$$

is bounded on $L^p(\mathbb{R}^2)$—where B is the unit ball in $\mathbb{R}^N$. Here $\vee$ denotes the inverse Fourier transform. This remarkable statement has several interesting features. First, the question of whether the infinite sequence of partial-summation operators converges is reduced by this reasoning to a question about a *single* linear operator (this statement is analogous to what the Hilbert transform says about norm convergence in the one-variable case). Second, that single operator is defined not on the torus $\mathbb{T} \times \mathbb{T}$ but rather on $\mathbb{R}^2$. In short, the final arbiter of the basic question that we wish to ask is a translation-invariant (convolution) operator on Euclidean space.

The operator $\mathcal{M}_B$ is called the "multiplier operator for the ball." We call it a "multiplier" because it involves multiplying the Fourier transform of f by an auxiliary function (in this case χ_B) and then taking the inverse Fourier transform. More generally, if m is a measurable function on $\mathbb{R}^N$, then the Fourier multiplier induced by m is the mapping

$$f \longmapsto \left[m \cdot \widehat{f} \right]^{\vee} \equiv \mathcal{M}_m f.$$

For fixed p, it is a matter of great interest to characterize those functions m with the property that $\mathcal{M}_m$ is a bounded operator on L^p (such an m would then be called a "Fourier multiplier for L^p" or just a "p-multiplier"). While there are many interesting results which give partial answers to the question (see [STE1], [STE2], and references therein), the only Lebesgue spaces for whose multipliers we have a complete characterization are L^2 and L^1. Here is the result for L^2:

Proposition 3.3.1 *Let μ be a measurable function on $\mathbb{R}^N$. Then μ is a Fourier multiplier for L^2 if and only if μ is essentially bounded (i.e., $\mu \in L^\infty$).*

Proof. For the sufficiency, suppose that μ is essentially bounded by M. If $f \in L^2$, then both $\widehat{f}$ and $\mu \widehat{f}$ lie in L^2. The Plancherel theorem tells us that

$$\|(2\pi)^{N/2} \mathcal{M}_\mu f\|_{L^2} = \|[\mathcal{M}_\mu f]\widehat{}\|_{L^2} = \|\mu \cdot \widehat{f}\|_{L^2} \le \|\mu\|_\infty \cdot \|\widehat{f}\|_2.$$

For the necessity, we wish to show (by Plancherel) that, if $\mu \notin L^\infty$, then $g \mapsto \mu \cdot g$ is not bounded on L^2. For $a > 0$ and $0 < R < \infty$, let

$$E_{a,R} = \{x : |\mu(x)| > a, |x| < R\}.$$

Then $m(E_{a,R}) < \infty$ since the measure is certainly bounded above by $(2R)^N$. And for any particular a we have that $m(E_{a,R}) > 0$ when R is large enough since $\mu \notin L^\infty$. Obviously $\|\mu \cdot \chi_{E_{a,R}}\|_{L^2} \ge a\|\chi_{E_{a,R}}\|_{L^2}$.

Since a can be arbitrarily large, we conclude that the operator of multiplication by μ is unbounded. $\square$

[We note in passing that μ is a Fourier multiplier for L^1 if and only if μ is the Fourier transform of a measure; this is surely a characterization, but more indirect than that for L^2 Fourier multipliers. See [STG1] for the details.]

Fourier multipliers are essentially self-dual. The argument that we now present is similar to the proof of Lemma 1.6.15 but is well worth repeating. Suppose that m is a Fourier multiplier for some L^p with $1 < p < \infty$. Let $p' = p/(p-1)$ be the conjugate exponent for p. Then, for f a testing function (see Appendix I),

$$\|\mathcal{M}_m f\|_{L^{p'}} = \sup_{\substack{\phi \in L^p \\ \|\phi\|_{L^p}=1}} \left| \int (\mathcal{M}_m f) \cdot \widetilde{\phi} \, dx \right|$$

$$= \sup_{\substack{\phi \in L^p \\ \|\phi\|_{L^p}=1}} \left| \int [m \cdot \widehat{f}]^{\vee} \cdot \widetilde{\phi} \, dx \right|$$

$$\overset{\text{(Proposition 2.2.7)}}{=} \sup_{\substack{\phi \in L^p \\ \|\phi\|_{L^p}=1}} \left| \int m \cdot \widehat{f} \cdot \widehat{\phi} \, dx \right|$$

$$= \sup_{\substack{\phi \in L^p \\ \|\phi\|_{L^p}=1}} \left| \int \widehat{f} \cdot [m \cdot \widehat{\phi}] \, dx \right|$$

$$\overset{\text{(Proposition 2.2.7)}}{=} \sup_{\substack{\phi \in L^p \\ \|\phi\|_{L^p}=1}} \left| \int \widetilde{f} [m \cdot \widehat{\phi}]^{\vee} dx \right|$$

$$\leq \sup_{\substack{\phi \in L^p \\ \|\phi\|_{L^p}=1}} \|f\|_{L^{p'}} \cdot \|\mathcal{M}_m \phi\|_{L^p}$$

$$\leq C' \cdot \sup_{\substack{\phi \in L^p \\ \|\phi\|_{L^p}=1}} \|f\|_{L^{p'}} \cdot \|\phi\|_{L^p}$$

$$= C' \cdot \|f\|_{L^{p'}}.$$

A simple limiting argument now enables passage from the testing functions to all of $L^{p'}$. Thus we see that if m is a Fourier multiplier for L^p, then m is also a Fourier multiplier for $L^{p'}$. [A slightly different argument works when $p = 1$, $p' = \infty$.] By an interpolation theorem of M. Riesz and G. O. Thorin (see Appendix III or [STG1] or [FOL1]), we may conclude that m is a Fourier multiplier for L^2. Thus, by Proposition 3.3.1, we may conclude that m is essentially bounded.

The upshot of the preceding discussion is that the only functions that we consider as candidates for Fourier multipliers are measurable, essentially bounded functions. And in this section we will be primarily concerned with functions m that are the characteristic functions of geometric sets in the plane. In particular, we will be interested in geometric sets that are models for the summation regions for indices that we wish to study (as an example, in the first paragraph we discussed the characteristic function of the unit disc as a Fourier multiplier, because it relates to spherical summability of double Fourier series). In the remainder of this section, we shall explain in some detail the connection between Fourier multipliers and summation of multiple Fourier series.

Our first lemma is this:

Lemma 3.3.2 *Suppose that f is a continuous function on $\mathbb{R}^N$. If f is 2π-periodic in each variable separately, then*

$$\lim_{\epsilon \to 0} (2\pi)^{-N/2} \epsilon^{N/2} \int_{\mathbb{R}^N} f(x) e^{-\epsilon |x|^2/2} \, dx = \frac{1}{(2\pi)^N} \int_{Q_N} f(x) \, dx \tag{3.3.2.1}$$

where Q_N is the fundamental region $Q_N \equiv \{(x_1, \dots, x_N) : -\pi \leq x_j < \pi\}$.

Proof. First the heuristic: Let $\Lambda = 2\pi \mathbb{Z} \times \cdots \times 2\pi \mathbb{Z}$. Then the left side of (3.3.2.1) is

$$\lim_{\epsilon \to 0^+} \sum_{m \in \Lambda} (2\pi)^N \left[\frac{1}{(2\pi)^N} \int_{m+Q_N} f(x) [(2\pi)^{-N/2} \epsilon^{N/2} e^{-\epsilon |x|^2/2}] \right] dx. \tag{3.3.2.2}$$

Note that the function

$$\eta(x) = (2\pi)^{-N/2} e^{-|x|^2/2}$$

is positive and satisfies $\int \eta(x)\, dx = 1$ (see Corollary 2.3.3). By change of variables, the functions

$$\eta_{\sqrt{\epsilon}}(x) \equiv (2\pi)^{-N/2} \epsilon^{N/2} e^{-\epsilon|x|^2/2}$$

satisfy $\int \eta_{\sqrt{\epsilon}}\, dx = 1$. However, when $\epsilon > 0$ is small, the function $\eta_{\sqrt{\epsilon}}$ has its mass more and more nearly uniformly distributed across $\mathbb{R}^N$. More precisely, if $\mathcal{K} \equiv [-M, M] \times \cdots \times [-M, M]$ is a compact subset of $\mathbb{R}^N$ then $\eta_{\sqrt{\epsilon}} \approx (2\pi)^{-N/2} \cdot \epsilon^{N/2}$ uniformly on $\mathcal{K}$.

Thus we see from (3.3.2.2) that the left-most integral in (3.3.2.1) in the statement of the lemma amounts (approximately) to averaging the terms

$$\frac{1}{(2\pi)^N} \int_{m+Q_N} f(x)\, dx,$$

with total mass 1. Since f is 2π-periodic, the result follows.

To make this argument precise, we note that the lemma is obviously true when $f(x) = e^{ij \cdot x}$, $j = (j_1, \ldots, j_N) \in \mathbb{Z}^N$, by considering the Fourier transform of the Gaussian—see (2.3.8). But, by the Stone-Weierstrass theorem (Appendix VIII), linear combinations of the functions $e^{ij \cdot x}$ are uniformly dense in the continuous, multiply periodic functions. The result follows. $\square$

The next lemma introduces the summability method that enables us to pass between $\mathbb{R}^N$ and $\mathbb{T}^N$.

Lemma 3.3.3 *Let P, Q be trigonometric polynomials. Let S be an L^p multiplier operator on $\mathbb{R}^N$, $1 < p < 2$, with multiplier s. Assume that the function s is continuous at each point $j \in \mathbb{Z}^N$. Suppose that $f(x) = \sum a_j e^{ij \cdot x}$ is a trigonometric polynomial. Set*

$$\widetilde{S} f(x) = \sum_{j \in \mathbb{Z}^N} s(-j) a_j e^{ij \cdot x}.$$

Let $\alpha, \beta > 0$ satisfy $\alpha + \beta = 1$. Then

$$\lim_{\epsilon \to 0} \left(\frac{2\pi}{\alpha\beta\epsilon} \right)^{N/2} \int_{\mathbb{R}^N} \left[S(P \cdot \eta_{\sqrt{\epsilon\alpha}})(x) \right] \left[\overline{Q(x) \cdot \eta_{\sqrt{\epsilon\beta}}(x)} \right] dx$$

$$= (2\pi)^{-N} \int_{Q_N} \left[(\widetilde{S}P)(x) \right] \left[\overline{Q(x)} \right] dx. \qquad (3.3.3.1)$$

Proof. Linearity tells us that it suffices to verify the result when $P(x) = e^{ij \cdot x}$ and $Q(x) = e^{ik \cdot x}$. We apply the polarized form of Plancherel's theorem (Proposition 2.3.18) to the left side of (3.3.3.1). Thus we have

$$\left(\frac{2\pi}{\alpha\beta\epsilon} \right)^{N/2} \int_{\mathbb{R}^N} \left[S(P \cdot \eta_{\sqrt{\epsilon\alpha}})(x) \right] \left[\overline{Q(x) \cdot \eta_{\sqrt{\epsilon\beta}}(x)} \right] dx$$

$$= \left(\frac{2\pi}{\alpha\beta\epsilon} \right)^{N/2} (2\pi)^{-N} \int_{\mathbb{R}^N} s(\xi)\phi(\xi) \cdot \overline{\psi(\xi)} \, d\xi; \qquad (3.3.3.2)$$

Here ϕ denotes the Fourier transform of

$$\left(P \cdot \eta_{\sqrt{\epsilon\alpha}} \right)(x) = e^{ij \cdot x} (2\pi)^{-N/2} (\epsilon\alpha)^{N/2} e^{-\epsilon\alpha|x|^2/2}$$

and ψ denotes the Fourier transform of

$$\left(Q \cdot \eta_{\sqrt{\epsilon\beta}} \right)(x) = e^{ik \cdot x} (2\pi)^{-N/2} (\epsilon\beta)^{N/2} e^{-\epsilon\beta|x|^2/2}.$$

By the calculations we did in connection with the theory of Gaussian summation,

$$\phi(\xi) = e^{-|\xi+j|^2/2\epsilon\alpha}$$

and

$$\psi(\xi) = e^{-|\xi+k|^2/2\epsilon\beta}.$$

Now the proof divides into two cases, according to whether $j \neq k$ or $j = k$.

Case 1: If $j \neq k$ then $|j - k| \geq 1$. Since s is a Fourier multiplier, we can be sure that s is bounded by some constant C. Thus the right side of equation (3.3.3.2) is majorized by

$$\left(\frac{2\pi}{\alpha\beta\epsilon}\right)^{N/2}(2\pi)^{-N}C\cdot\int_{\mathbb{R}^N}e^{-|\xi+j|^2/2\alpha\epsilon}e^{-|\xi+k|^2/2\beta\epsilon}\,d\xi$$

$$\le C'\epsilon^{-N/2}\left[\int_{|\xi+j|\ge 1/2}+\int_{|\xi+k|\ge 1/2}\right]$$

$$\equiv I+II.$$

On the domain of I the term

$$e^{-|\xi+j|^2/2\alpha\epsilon}\le e^{-1/8\alpha\epsilon}$$

tends uniformly to 0 as $\epsilon \to 0$, while the complementary factor $e^{-|\xi+k|^2/2\beta\epsilon}$ is majorized by the single L^1 function $e^{-|\xi+k|^2/2\beta}$ as soon as $\epsilon < 1$.

We therefore see that $I \to 0$ as $\epsilon \to 0$. In the same fashion, $II \to 0$ as $\epsilon \to 0$. By comparison, the right side of (3.3.3.1) is

$$(2\pi)^{-N}\int_{Q_N}s(-j)e^{ij\cdot x}e^{-ik\cdot x}\,dx=0$$

since $j \ne k$. This completes the argument when $j \ne k$.

Case 2: Now we consider the case $j = k$. Since

$$\frac{1}{\alpha}+\frac{1}{\beta}=\frac{1}{\alpha\beta},$$

the left side of (3.3.3.1) is the Gauss-Weierstrass integral (as in (3.3.3.2))

$$\left(\frac{2\pi}{\alpha\beta\epsilon}\right)^{N/2}(2\pi)^{-N}\int_{\mathbb{R}^N}s(\xi)\phi(\xi)\overline{\psi(\xi)}\,d\xi$$

$$=\left(\frac{2\pi}{\alpha\beta\epsilon}\right)^{N/2}(2\pi)^{-N}\int_{\mathbb{R}^N}s(\xi)e^{-|\xi+j|^2/2\alpha\beta\epsilon}\,d\xi$$

$$=\int_{\mathbb{R}^N}s(\xi)G_{\alpha\beta\epsilon}(j+\xi)\,d\xi.$$

Of course this integral will converge to $s(-j)$ as $\epsilon \to 0^+$, since $-j$ is a point of continuity of s. The right side of (3.3.3.1) clearly has this same value. Therefore the argument is complete. $\qquad\square$

The next is our principal result for this section:

Theorem 3.3.4 *Let* $1 < p < \infty$ *and suppose that* S *is an* L^p *multiplier operator for* $\mathbb{R}^N$. *Let* s *be the multiplier associated to* S, *and assume that* s *is continuous at each point of* $\mathbb{Z}^N$. *Then there is a unique periodized operator* $\widetilde{S}$ *given by*

$$\widetilde{S}f \sim \sum_{j \in \mathbb{Z}^N} s(-j)\widehat{f}(j)e^{ij\cdot x}, \quad f \in L^p(\mathbb{T}^N),$$

such that $\widetilde{S}$ *is a Fourier multiplier for* $L^p(\mathbb{T}^N)$. *Furthermore,*

$$\|\widetilde{S}\|_{(L^p(\mathbb{T}^N), L^p(\mathbb{T}^N))} \le \|S\|_{(L^p(\mathbb{R}^N), L^p(\mathbb{R}^N))}.$$

Proof. Let $p' = p/(p-1)$ be the conjugate exponent. We first obtain an *a priori* bound for the operator $\widetilde{S}$ when it is restricted to trigonometric polynomials. Let P and Q be trigonometric polynomials.

With η_ϵ as above, we consider the expression

$$\left(\frac{2\pi}{\alpha\beta\epsilon}\right)^{N/2} \int_{\mathbb{R}^N} \left[S(P \cdot \eta_{\sqrt{\epsilon\alpha}})(x)\right]\left[\,\overline{Q(x) \cdot \eta_{\sqrt{\epsilon\beta}}(x)}\,\right] dx, \quad (3.3.4.1)$$

with $\alpha = 1/p$ and $\beta = 1/p'$. We let $\epsilon \to 0$. The result, by Lemma 3.3.3, is

$$(2\pi)^{-N} \int_{Q_N} (\widetilde{S}P)(x)\overline{Q}(x)\, dx. \quad (3.3.4.2)$$

On the other hand, formula (3.3.4.1) is majorized by

$$\left(\frac{2\pi}{\alpha\beta\epsilon}\right)^{N/2} \|S\|_{\mathrm{op}} \|P\eta_{\sqrt{\epsilon\alpha}}\|_{L^p(\mathbb{R}^N)} \|Q\eta_{\sqrt{\epsilon\beta}}\|_{L^{p'}(\mathbb{R}^N)}$$

$$= \left(\frac{2\pi}{\alpha\beta\epsilon}\right)^{N/2} \|S\|_{\text{op}}$$

$$\times \left[\int_{\mathbb{R}^N} |P(x)|^p [(2\pi)^{-N/2}(\epsilon\alpha)^{N/2}]^p e^{-\epsilon\alpha p|x|^2/2}\, dx\right]^{1/p}$$

$$\times \left[\int_{\mathbb{R}^N} |Q(x)|^{p'} [(2\pi)^{-N/2}(\epsilon\beta)^{N/2}]^{p'} e^{-\epsilon\beta p'|x|^2/2}\, dx\right]^{1/p'}$$

$$= \left(\frac{2\pi}{\alpha\beta\epsilon}\right)^{N/2} \|S\|_{\text{op}}$$

$$\times \left[\int_{\mathbb{R}^N} |P(x)|^p [(2\pi)^{-N/2}(\epsilon\alpha)^{N/2}]^p e^{-\epsilon|x|^2/2}\, dx\right]^{1/p}$$

$$\times \left[\int_{\mathbb{R}^N} |Q(x)|^{p'} [(2\pi)^{-N/2}(\epsilon\beta)^{N/2}]^{p'} e^{-\epsilon|x|^2/2}\, dx\right]^{1/p'}$$

$$= \|S\|_{\text{op}} \left[\int_{\mathbb{R}^N} |P(x)|^p (2\pi)^{-N/2}\epsilon^{N/2} e^{-\epsilon|x|^2/2}\, dx\right]^{1/p}$$

$$\times \left[\int_{\mathbb{R}^N} |Q(x)|^{p'} (2\pi)^{-N/2}\epsilon^{N/2} e^{-\epsilon|x|^2/2}\, dx\right]^{1/p'}$$

$$= \|S\|_{\text{op}} \left[\int_{\mathbb{R}^N} |P(x)|^p \eta_{\sqrt{\epsilon}}(x)\, dx\right]^{1/p}$$

$$\times \left[\int_{\mathbb{R}^N} |Q(x)|^{p'} \eta_{\sqrt{\epsilon}}(x)\, dx\right]^{1/p'}.$$

As $\epsilon \to 0^+$, this expression tends (by Lemma 3.3.2) to

$$\|S\|_{\text{op}} \left[\frac{1}{(2\pi)^N} \int_{Q_N} |P(x)|^p\, dx\right]^{1/p} \cdot \left[\frac{1}{(2\pi)^N} \int_{Q_N} |Q(x)|^{p'}\, dx\right]^{1/p'}.$$

Now we combine this last majorization with (3.3.4.2) to obtain

$$\left|\int_{Q_N} (\widetilde{S}P)(x)\overline{Q(x)}\, dx\right| \le \|S\|_{\text{op}}\|P\|_{L^p(Q_N)}\|Q\|_{L^{p'}(Q_N)}.$$

If we now take the supremum over all trigonometric polynomials Q such that $\|Q\|_{L^{p'}} \leq 1$, then we obtain (using Theorem 0.3.9)

$$\|\widetilde{S}P\|_{L^p(\mathbb{T}^N)} \leq \|S\|_{\mathrm{op}}\|P\|_{L^p(Q_N)}, \quad \text{any trigonometric polynomial } P.$$

Since the trigonometric polynomials are dense in L^p, $1 < p < \infty$, we see that there is a unique bounded extension of $\widetilde{S}$ to all of $L^p(\mathbb{T}^N)$.

□

Some technical modifications of the preceding proof give the following result, which we shall see to be useful in practice:

Corollary 3.3.5 *The theorem is still valid if instead of continuity of the multiplier we assume that, at each $j \in \mathbb{Z}^N$, we have*

$$\lim_{\epsilon \to 0} \epsilon^{-N} \int_{|t| \leq \epsilon} |s(j - t) - s(j)|\, dt = 0.$$

In effect, the corollary says that the theorem is valid as long as each lattice point is a Lebesgue point for the Fourier multiplier (see 0.2.16, 0.2.15 for Lebesgue's theorem).

Now the theorem gives us a way of passing from a Fourier multiplier for the Fourier transform on $\mathbb{R}^N$ to a multiplier for multiple Fourier series, and hence to a summation method for multiple Fourier series. It is of both philosophical and practical utility to have the following converse (which we shall not prove). Before we formulate this converse, note that it is unlikely that the following reasoning would work: Assume that s on $\mathbb{R}^N$ is a continuous function, and that the restriction of s to the lattice points is a multiplier on $\mathbb{T}^N$; then s is a multiplier on $\mathbb{R}^N$. The reason that this is an improbable implication is that it allows s to be virtually arbitrary at points away from the lattice. To address this logical issue, we enlist the dilations of Euclidean space as an aid (see [DAY], [STG1]):

Theorem 3.3.6 *Suppose that s is a continuous, complex-valued function on $\mathbb{R}^N$ and assume that, for each $\epsilon > 0$, the operators given by*

$$\widetilde{S}_\epsilon f(x) \sim \sum_{j \in \mathbb{Z}^N} s(-\epsilon j)\, \widehat{f}(j) e^{ij \cdot x}$$

are uniformly bounded on $L^p(\mathbb{T}^N)$ (that is, with a uniform bound, independent of ϵ, on their operator norms) for $1 < p < \infty$. Then s is a Fourier multiplier on $L^p(\mathbb{R}^N)$. Furthermore,

$$\|\mathcal{M}_s\| \leq \sup_{\epsilon > 0} \|\widetilde{S}_\epsilon\|,$$

where $\mathcal{M}_s$ is the linear operator on $L^p(\mathbb{R}^N)$ corresponding to the multiplier s.

3.4   Applications of the Fourier Multiplier Theorems to Summation of Multiple Trigonometric Series

Let us think about how to apply the two theorems of the preceding section to the problem of summing multiple Fourier series by the various methods introduced in Section 3.1. We will concentrate on *norm summation*.

First let us consider square or cubical summation. Suppose that we let $s = \chi_Q$, where Q is the unit square/cube in Euclidean N-space (i.e., the square/cube with center 0, side length 2, and sides parallel to the axes) and χ denotes the characteristic function. Fix an index $1 < p < \infty$. Assume that it is known that s is a Fourier multiplier on $L^p(\mathbb{R}^N)$. We would like to apply Theorem 3.3.4. Note that we cannot apply the theorem directly because χ_Q is not continuous at each lattice point. The matter can be handled in two ways:

(i) We may dilate Q by an arbitrarily small amount so that the resulting cube $\widetilde{Q}$ has the property that any lattice point is either strictly interior to the *open* cube $\widetilde{Q}$ or else is strictly exterior to the *closed* cube $\overline{\widetilde{Q}}$. As a result, the hypotheses of the corollary to the theorem will certainly be satisfied. And we may conclude that the periodized operator

$$f \longmapsto \sum_{j \in \mathbb{Z}^N} \chi_{\widetilde{Q}}(-j)\widehat{f}(j)e^{ij \cdot x}$$

is bounded on $L^p(\mathbb{T}^N)$. The bound on this operator can be taken to be independent of the (arbitrarily small) perturbation $\widetilde{Q}$ of Q. As a result, we may conclude that

$$\mathcal{M}_s : f \longmapsto \sum_{j \in \mathbb{Z}^N} \chi_Q(-j)\widehat{f}(j)e^{ij \cdot x}$$

(we are defining $\mathcal{M}_s$ by this formula) is bounded on $L^p(\mathbb{T}^N)$.

(ii) We may invoke Corollary 3.3.5 directly, noting that if any lattice point falls on the boundary of Q, then the characteristic function differentiates (in the sense of Lebesgue) at that lattice point to one of the values $1/2$ (if the point is in a face of Q) or $1/2^{N-1}$ (if the point is in the interior of an edge of Q) or $1/2^N$ (if the point is at a corner of Q). Thus the corollary gives rise to a summation method for Fourier series that involves multiplying certain coefficients coming from the boundary of the coefficient region by either $1/2$ or $1/2^{N-1}$ or $1/2^N$. But then the Cantor-Lebesgue lemma (see [COO] and [CON]) shows that the contribution from those boundary terms is negligible.

Lemma 3.4.1 *Fix $1 < p < \infty$. Let Q_R be the R-fold dilate of the unit cube Q. If $\mathcal{M}_Q$ is bounded on $L^p(\mathbb{R}^N)$, then the operator*

$$\mathcal{M}_{Q_R} : f \longmapsto \left[\chi_{Q_R}\widehat{f}\right]^{\vee}$$

is bounded on $L^p(\mathbb{R}^N)$, independent of R.

Corollary 3.4.2 *If $\mathcal{M}_Q$ is bounded on $L^p(\mathbb{R}^N)$ for some $1 < p < \infty$, then the periodized operator*

$$\mathcal{M}_{Q_R} : f \longmapsto \sum_{j \in \mathbb{Z}^N} \chi_{Q_R}(-j)\widehat{f}(j)e^{ij \cdot x}$$

is bounded on $L^p(\mathbb{T}^N)$, independent of R.

We shall prove the lemma in detail. The corollary follows, of course, by an application of Theorem 3.3.4, as above.

Proof of the Lemma. So that we need not worry about convergence of the integrals, we take f to be in the Schwartz class (Appendix II)— which is certainly dense in L^p. We calculate that

$$\begin{aligned}
\left[\chi_{Q_R}\widehat{f}\right]^{\vee}(x) &= (2\pi)^{-N} \int \chi_{Q_R}(\xi)\widehat{f}(\xi)e^{-ix \cdot \xi} \, d\xi \\
&= R^N (2\pi)^{-N} \int \chi_{Q_R}(R\xi)\widehat{f}(R\xi)e^{-ix \cdot R\xi} \, d\xi \\
&= R^N (2\pi)^{-N} \int \chi_Q(\xi)\widehat{\alpha^R f}(\xi)e^{-ix \cdot R\xi} \, d\xi \\
&= R^N \left[\mathcal{M}_Q(\alpha^R f)\right](Rx).
\end{aligned}$$

We have just calculated $\mathcal{M}_{Q_R} f$. Now we need to calculate its L^p norm:

$$\begin{aligned}
\|\mathcal{M}_{Q_R} f\|_{L^p} &= R^N \left[\int |\mathcal{M}_Q[\alpha^R f](Rx)|^p \, dx\right]^{1/p} \\
&= R^N \cdot R^{-N/p} \left[\int |\mathcal{M}_Q[\alpha^R f](x)|^p \, dx\right]^{1/p} \\
&\leq C \cdot R^N \cdot R^{-N/p} \|\alpha^R f\|_{L^p} \\
&= C \cdot R^N \cdot R^{-N/p} \left[\int |R^{-N} f(R^{-1}x)|^p \, dx\right]^{1/p} \\
&= C \cdot R^N \cdot R^{-N/p} \left[\int R^N \cdot R^{-Np} |f(x)|^p \, dx\right]^{1/p} \\
&= C \cdot \left[\int |f(x)|^p \, dx\right]^{1/p}.
\end{aligned}$$

Thus we have a bound on the L^p norm of $\mathcal{M}_{Q_R} f$, depending on f of course but independent of R. That completes the proof. $\qquad\square$

In any event, the arguments just presented show that the boundedness on $L^p(\mathbb{R}^N)$ of the single Fourier multiplier coming from the unit cube in $\mathbb{R}^N$ would imply that cubic summation for N-dimensional Fourier series is valid in $L^p(\mathbb{T}^N)$. The converse holds as well, by Theorem 3.3.6. [Again, the hypothesis of Theorem 3.3.6 can be weakened to supposing that each lattice point is a Lebesgue point for s—just as we have discussed in connection with Corollary 3.3.5.]

In view of the preceding discussion, we now analyze the multiplier problem for the characteristic function of the unit cube in $\mathbb{R}^N$. For convenience, we treat only the case $N = 2$. We will show that that multiplier is easily dispatched by reducing the question to the study of a variant of the Hilbert transform on $\mathbb{T}$ that we discussed in Chapter 1. Namely, the *Hilbert transform* on $\mathbb{R}$ is the operator T defined by the multiplier $-i \operatorname{sgn} \xi$:

$$(H\phi)\widehat{}(\xi) = -i \operatorname{sgn}(\xi)\widehat{\phi}(\xi).$$

A calculation similar to those in Chapter 1 shows that H can also be expressed as

$$H\phi(x) = \text{P.V.} \frac{1}{\pi} \int_{-\infty}^{\infty} \frac{\phi(x-t)}{t}\, dt;$$

moreover, the analogue of Theorem 1.6.11 is valid (see also the discussion before and after that theorem). To wit, H is bounded on L^p for $1 < p < \infty$. In turn, this result is actually a special case of results that we shall prove in Chapter 6 about classes of singular integral operators.

The nub of our discussion is the following basic result:

Theorem 3.4.3 *Let P be a point of $\mathbb{R}^2$, $\mathbf{v} \in \mathbb{R}^2$ be a unit vector, and set*

$$E_{\mathbf{v}} = \{x \in \mathbb{R}^2 : (x - P) \cdot \mathbf{v} \geq 0\}.$$

Then the operator

$$f \longmapsto \left(\chi_{E_\mathbf{v}} \cdot \widehat{f} \right)^{\vee}$$

is bounded on L^p, $1 < p < \infty$.

Proof. As just noted, the 1-dimensional Hilbert transform

$$\phi \mapsto \left(-i \operatorname{sgn} \xi \cdot \widehat{\phi} \right)^{\vee}$$

is bounded on $L^p(\mathbb{R})$, $1 < p < \infty$.[1] As we also saw in Chapter 2, it is often useful to consider, instead of the Hilbert transform H, the operator $M = \frac{1}{2}(I + iH)$ because it has the very simple Fourier multiplier $m = \chi_{[0,\infty)}$. This we now do.

We now express the multiplier for a half-space as an amalgam of multipliers for the half-line—using Fubini's theorem. As already noted, we restrict attention to dimension 2.

After composition with a rotation and a translation, we may assume that $P = 0$ and that the vector $\mathbf{v}$ is the vector $(0, 1)$. Let us drop the subscript and denote the corresponding half-space by E.

Fix $1 < p < \infty$. Since the Schwartz functions are dense in $L^p(\mathbb{R}^2)$ (see Appendix I), it suffices for us to perform the estimates for f a Schwartz function. For almost every $x_1 \in \mathbb{R}^1$, the function $f_{x_1}(x_2) \equiv f(x_1, x_2)$ is certainly in $L^p(\mathbb{R}^1)$ and, by Fubini's theorem,

$$\int_{x_1 \in \mathbb{R}} \| f_{x_1} \|_{L^p(\mathbb{R})}^p dx_1 = \| f \|_{L^p(\mathbb{R}^2)}^p.$$

Now we have

$$\left(\chi_E \widehat{f} \right)^{\vee}(x_1, x_2) = (2\pi)^{-2} \int_0^\infty \int_\mathbb{R} \int_\mathbb{R} \int_\mathbb{R} f(t_1, t_2)$$
$$\times e^{i\xi_1 t_1} e^{i\xi_2 t_2} dt_1 dt_2 e^{-i\xi_1 x_1} e^{-i\xi_2 x_2} d\xi_1 d\xi_2.$$

---

[1] We discussed this fact in some detail in Chapter 1, and it will be a corollary of the Calderón-Zygmund theorem in Chapter 6.

The two inside integrals give rise to a Schwartz function, so all integrals converge absolutely. By Fubini's theorem, the last line equals

$$\frac{1}{2\pi}\int_{\mathbb{R}}\int_{\mathbb{R}}\left[\frac{1}{2\pi}\int_0^\infty\left[\int_{\mathbb{R}}f(t_1,t_2)e^{i\xi_2 t_2}dt_2\right]e^{-i\xi_2 x_2}d\xi_2\right]$$
$$\times e^{i\xi_1 t_1}dt_1 e^{-i\xi_1 x_1}d\xi_1$$
$$=\frac{1}{2\pi}\int_{\mathbb{R}}\int_{\mathbb{R}}M(f_{t_1})(x_2)e^{it_1\xi_1}dt_1 e^{-i\xi_1 x_1}d\xi_1. \qquad (3.4.3.1)$$

Since, for almost every t_1, $f_{t_1}\in L^p(\mathbb{R})$, we see that Mf_{t_1} makes sense for each t_1. Also

$$\int_{t_1\in\mathbb{R}}\|Mf_{t_1}(\cdot)\|^p_{L^p(\mathbb{R})}dt_1 \le C\int_{t_1\in\mathbb{R}}\|f_{t_1}(\cdot)\|^p_{L^p(\mathbb{R})}dt_1$$
$$=C\|f\|^p_{L^p(\mathbb{R}^2)}. \qquad (3.4.3.2)$$

Hence we see that $Mf_{t_1}(x_2)$ is in $L^p(\mathbb{R}^2)$ as a function of the variables (t_1,x_2). In particular, for almost every x_2, the function

$$t_1\longmapsto F_{x_2}(t_1)\equiv M(f_{t_1})(x_2)$$

lies in $L^p(\mathbb{R})$.

Also,

$$\int_{x_2\in\mathbb{R}}\|F_{x_2}(\cdot)\|^p_{L^p(\mathbb{R})}dx_2=\int_{t_1\in\mathbb{R}}\|M(f_{t_1})(\cdot)\|^p_{L^p(\mathbb{R})}dt_1\le C\cdot\|f\|^p_{L^p(\mathbb{R}^2)}$$

by (3.4.3.2).

In summary, we may rewrite the right-hand side of (3.4.3.1) as

$$\lim_{\epsilon\to 0}\int_{\mathbb{R}}\frac{1}{2\pi}\int_{\mathbb{R}}F_{x_2}(t_1)e^{it_1\xi_1}dt_1 e^{-i\xi_1 x_1}e^{-\epsilon|\xi_1|^2}d\xi_1$$

by Gauss-Weierstrass summation. But this equals

$$\lim_{\epsilon\to 0}\frac{1}{2\pi}\int_{\mathbb{R}}\check{F}_{x_2}(-\xi_1)e^{-i\xi_1 x_1}e^{-\epsilon|\xi_1|^2}d\xi_1=F_{x_2}(x_1),$$

for almost every x_1. We have already noted that the latter function has $L^p(\mathbb{R}^2)$ norm dominated by $C\|f\|_{L^p(\mathbb{R}^2)}$. The proof of the theorem is therefore complete.   □

Theorem 3.4.4 *The method of square summation is valid for double trigonometric series in the L^p norm, $1 < p < \infty$.*

Proof. Let

$$E_1 \equiv \{(x, y) \in \mathbb{R}^2 : (-1, 0) \cdot [(x, y) - (1, 0)] \geq 0\},$$

$$E_2 \equiv \{(x, y) \in \mathbb{R}^2 : (1, 0) \cdot [(x, y) - (-1, 0)] \geq 0\},$$

$$E_3 \equiv \{(x, y) \in \mathbb{R}^2 : (0, -1) \cdot [(x, y) - (0, 1)] \geq 0\},$$

$$E_4 \equiv \{(x, y) \in \mathbb{R}^2 : (0, 1) \cdot [(x, y) - (0, -1)] \geq 0\}.$$

Then E_1, E_2, E_3, E_4 are four half-planes whose common intersection is the unit square $Q = \{(x, y) : |x| \leq 1, |y| \leq 1\}$ in $\mathbb{R}^2$. Let T_j be the multiplier operator associated to χ_{E_j}, that is,

$$T_j : f \longmapsto (\chi_{E_j} \cdot \widehat{f})^{\vee}.$$

Then $T_1 \circ T_2 \circ T_3 \circ T_4$ is the multiplier operator associated to the closed unit square.

We know from Theorem 3.4.3 that each T_j is bounded on L^p, $1 < p < \infty$. As a result, $T_1 \circ T_2 \circ T_3 \circ T_4$ is certainly bounded on L^p for the same range of p. Therefore the multiplier operator associated to the unit square is bounded on L^p, $1 < p < \infty$. As a result, square summability is valid for double Fourier series, $1 < p < \infty$.   □

The reader may check for himself that similar arguments show that square summability is valid for Fourier series on $\mathbb{T}^N$, any $N = 2, 3, 4, \ldots$. More importantly, exactly the same arguments show that polygonal summability is valid on any $\mathbb{T}^N$, $1 < p < \infty$. More precisely, we have:

Theorem 3.4.5 *Let P be any closed, convex polygon (convex polyhedron) lying in $\mathbb{R}^2$ (resp., $\mathbb{R}^N$) and having nonempty interior.*

Then the Fourier multiplier $\mathcal{M}_{\chi_P}$ *is bounded on* L^p, $1 < p < \infty$. *Therefore polygonal summation with respect to this* P *is valid in* L^p, $1 < p < \infty$.

As an additional exercise, consider how to prove the theorem when the summation method is modeled on a *nonconvex* polygon.

Restricted rectangular convergence is quite a bit trickier than the types of polygonal convergence that we have been discussing. For instance, polygonal *pointwise almost everywhere* convergence is valid by a simple induction argument from the (quite difficult) 1-dimensional result (see [FEF5], [SJO2], [TEV]); but pointwise almost everywhere convergence fails for restricted rectangular convergence [FEF6]. An astonishing example of Fefferman (with modifications by Ash and Gluck) gives a continuous function on $\mathbb{T}^2$, with $\widehat{f}(j, k) = 0$ when either $j < 0$ or $k < 0$, and having everywhere uniformly bounded rectangular partial sums, but whose Fourier series diverges restrictedly rectangularly at every point. An authoritative treatment of some of the basic positive results about this method of convergence appears in [ZYG, p. 300 ff.].

We also mention that even square summability exhibits some surprising pathologies. The paper [KONS] describes some of the newest of these.

It is easy to see that the arguments presented for polygonal convergence will not apply to spherical summation. In 2 dimensions, the disc is certainly the intersection of infinitely many half-planes. But the methodology we have presented thus far would then represent the disc multiplier as the composition of *infinitely many* half-plane multipliers. Thus the bound on the operator would (most likely) blow up.

Of course the preceding discussion shows that our *method of proof* will not suffice to show that the disc multiplier is a bounded operator on any L^p, $p \neq 2$. The possibility remains that there is some other method of proof that will give a favorable estimate. The arguments that we present in the next section show, in effect, that what seems to be going wrong with the suggested proof in the last paragraph is intrinsic: the disc multiplier is unbounded on any L^p, $p \neq 2$.

3.5   The Multiplier Problem for the Ball

We begin with a detailed discussion of a problem from the geometry of the plane that has no apparent connection with harmonic analysis. It was an astonishing discovery of C. Fefferman, in 1971, that the
construction we shall describe has profound connections with multi-
dimensional harmonic analysis. [Note that the construction collapses
to nothing in 1 dimension, as does the harmonic analysis problem being considered.] In fact our current understanding of Euclidean harmonic analysis, due to C. Fefferman, E. M. Stein, G. Weiss, and others,
reveals that virtually any deep question about multi-dimensional harmonic analysis relates to fundamental ideas about Euclidean geometry.

The genesis of the geometric problem just alluded to comes from
a nineteenth-century Japanese scientist named S. Kakeya. He posed
the following question (this is its classical formulation): Dip a sewing
needle into a bottle of ink. Now place it on a piece of paper. Endeavor to
move the needle on the piece of paper so as to **(i)** reverse the positions
of the two ends, and **(ii)** leave behind an ink blot of smallest possible
area.

This question baffled mathematicians for decades. It was finally
solved by A. Besicovitch in 1928 (see [BES]).

Here are some preliminary thoughts about the question:

1. First suppose that the needle has length 2, and is of infinitesimal
   thickness. Let D be a closed disc in the plane of radius 1 (see Figure 1). Of course we may insert the needle inside D as shown in
   Figure 2. Furthermore, the needle may be rotated so as to switch
   its ends, while keeping the needle inside the disc all the while: Figure 3. In the language of the Kakeya problem: If we let k be the
   infimum of areas of all ink blots that can be left behind by a needle
   as specified by Kakeya then

   $$k \leq \text{area of unit disc} = \pi \cdot 1^2 = \pi.$$

2. To improve on the estimate in **1**, consider Figure 4. Pin down one
   end of the needle, and swing the other end through an angle of $\pi/3$

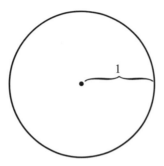

Figure 1.   A closed disc in the plane.

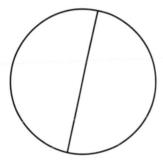

Figure 2.   A needle in the disc.

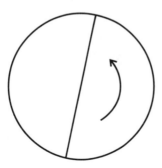

Figure 3.   Rotating the needle.

Figure 4.   Improved construction of a Kakeya set.

radians. This leaves behind an ink blot as shown. Now pin down the free end, and rotate the first end through an angle of $\pi/3$ radians. See Figure 5. Now repeat the rotation process a third time. The resulting ink blot is shown in Figure 6. Clearly, by construction, this is a Kakeya blot. We invite the reader to verify that the *area* of this blot is $2\pi - 2\sqrt{3}$, a number which is strictly less than π—the area of the blot in the last example.

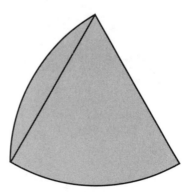

Figure 5.   Second step in the improved construction.

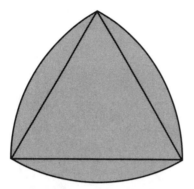

Figure 6.   Final step in the improved construction.

For many years, mathematicians thought that the blot described in this second example was in fact the solution to the Kakeya needle problem, that is, that this was the smallest blot possible.

Unlike some solutions to mathematics problems, which simply unleash an arsenal of weapons to confirm a predictable result, Besicovitch's solution to the Kakeya needle problem was a shock. For Besicovitch showed that, given any $\epsilon > 0$, there is a set $E \subseteq \mathbb{R}^2$ such that **(i)** the area of E is less than ϵ, and **(ii)** the needle may have its ends switched, in the manner specified in the statement of the Kakeya needle problem, in such a way that the needle remains in the set E at all times. In other words, it is possible to move the needle in the manner specified by Kakeya but so that the ink blot left behind has area less than ϵ, no matter how small $\epsilon > 0$.

We now present a modern variant of the Besicovitch construction which may be found in [CUN]. It turns out to be a bit more flexible than Besicovitch's original solution to the problem, and also a bit easier to understand. Afterwards, we will explain how it may be adapted to the study of spherical summation of multiple Fourier series.

We begin with a triangle T of base b and height h, as shown in Figure 7. The basic step in our construction is to *sprout* T to two new triangles, of height $h' > h$. Here is what happens during the sprouting

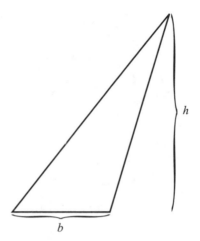

Figure 7.   The sprouting process.

process; refer to Figure 8 as you read along. We extend AC to a point A' at height h' and likewise we extend BC to a point B' at height h'. We connect A' and B' to the midpoint D of AB, as shown. The two triangles $T'_A \equiv \triangle AA'D$ and $T'_B \equiv \triangle BB'D$ are called *sprouts* from height h to h'. Of course we can later, if we wish, sprout the new triangles T'_A and T'_B to a new height $h'' > h'$. In fact we shall do this; more, we shall iterate our procedure some large positive number k times in order to obtain the desired estimates. The Kakeya set that we wish to construct will be, essentially, the union of all the sprouts.

Lemma 3.5.1 *With the triangle T and sprouts T'_A, T'_B specified as above, the area of $T'_A \cup T'_B$ is*

$$A_{T'_A \cup T'_B} = \frac{1}{2}bh + \frac{b(h' - h)^2}{2h' - h}.$$

Proof. Refer to Figure 9. Of course $\triangle B'DB$ is similar to $\triangle B'EC$, and $\triangle A'AD$ is similar to $\triangle A'CF$. Therefore

$$\frac{EC}{DB} = \frac{h' - h}{h'}$$

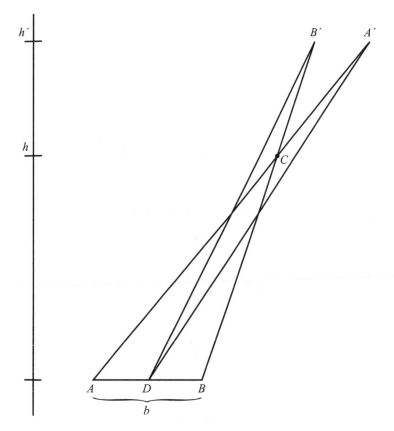

Figure 8. Details of the sprouting.

and

$$\frac{CF}{AD} = \frac{h' - h}{h'}.$$

As a result,

$$EC = \frac{h' - h}{h'} \cdot DB = \frac{h' - h}{h'} \cdot AD = CF.$$

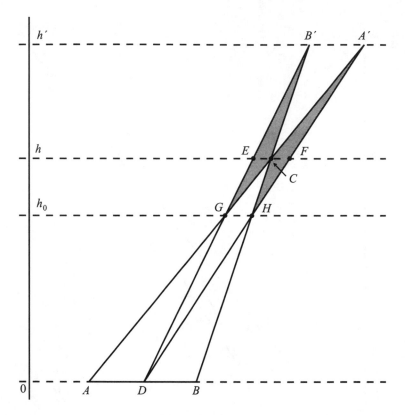

Figure 9. Area of the sprouts.

Thus $EC = CF$; $AD = DB = b/2$. As a result,

$$EC = \frac{h' - h}{h'} \cdot \frac{b}{2}$$

and

$$CF = \frac{h' - h}{h'} \cdot \frac{b}{2}.$$

Now notice that, by proportionality, the height of the point G above the base of T is just the same as the height of the point H above the base of T; we let that common height be h_0. Also note that

$$\frac{h - h_0}{h_0} = \frac{EC}{AD},$$

which we see by comparing similar triangles $\triangle AGD$ and $\triangle CGE$. Thus

$$h - h_0 = \frac{EC}{AD} \cdot h_0$$

or

$$h = h_0 \left(\frac{EC}{AD} + 1 \right)$$

and

$$h_0 = \frac{h}{EC/AD + 1} = \frac{h}{EC/DB + 1} = \frac{h}{\frac{h'-h}{h'} + 1} = \frac{hh'}{2h' - h}.$$

The area of $\triangle ACB \cup \triangle GB'C \cup \triangle HCA'$ is thus

$$\frac{1}{2}bh + \left[\frac{1}{2}EC(h' - h) + \frac{1}{2}EC(h - h_0) \right]$$

$$+ \left[\frac{1}{2}CF(h' - h) + \frac{1}{2}CF(h - h_0) \right]$$

$$= \frac{1}{2}bh + \frac{1}{2}EC(h' - h_0) + \frac{1}{2}CF(h' - h_0)$$

$$= \frac{1}{2}bh + \frac{1}{2} \left(\frac{h' - h}{h'} \cdot \frac{b}{2} \right) \cdot (h' - h_0)$$

$$+ \frac{1}{2} \left(\frac{h' - h}{h'} \cdot \frac{b}{2} \right) \cdot (h' - h_0)$$

$$= \frac{1}{2}bh + \left(\frac{h' - h}{h'} \cdot \frac{b}{2} \right) \left(h' - \frac{hh'}{2h' - h} \right)$$

$$= \frac{1}{2}bh + \frac{b(h' - h)^2}{2h' - h}.$$

This completes our calculation of the area of $A_{T'_A \cup T'_B}$. Observe in passing that each of the triangles $\triangle GB'C$ and $\triangle HCA'$ in the figure has area $b(h' - h)^2/[2(2h' - h)]$. $\square$

Now we wish to sprout the original triangle T multiple times and then estimate the area of the union of all the sprouts. We take an initial equilateral triangle having base $b_0 = \epsilon$ and height $h_0 = \sqrt{3}\epsilon/2$. Thus the initial triangle T_0 is an equilateral triangle of side ϵ.

Now we define a sequence of heights

$$h_0 = \frac{\sqrt{3}}{2}\epsilon$$

$$h_1 = \frac{\sqrt{3}}{2}\left(1 + \frac{1}{2}\right)\epsilon$$

$$h_2 = \frac{\sqrt{3}}{2}\left(1 + \frac{1}{2} + \frac{1}{3}\right)\epsilon$$

$$\cdots$$

$$h_{j-1} = \frac{\sqrt{3}}{2}\left(1 + \frac{1}{2} + \frac{1}{3} + \cdots + \frac{1}{j}\right)\epsilon.$$

We sprout T_0 from height h_0 to height h_1 to obtain two new triangles. The new triangles are called T_{01}, T_{02}. Note that every sprout of every generation has base that is a subsegment of AB. Thus the first generation sprouts T_{01} and T_{02} are certainly not disjoint. But the portions of T_{01} and T_{02} that protrude from $\triangle ABC$ have disjoint interiors (these are displayed as $\triangle GB'C$ and $\triangle HCA'$ in Figure 9). We sprout each of T_{01} and T_{02} to the new height h_2 to obtain four new triangles T_{011}, T_{012}, T_{021}, and T_{022}. We continue the process, obtaining at the jth sprouting a total of 2^j new triangles of height h_j and base $2^{-j}\epsilon$. For a large positive integer k to be fixed later, the Besicovitch set E (spanning an angle of $\pi/3$) that we seek will be the union of the 2^k triangles generated at the kth sprouting.

Now let us estimate the area of E. At the jth sprouting, the increment of area (i.e., the area actually added at that stage) is composed of 2^j triangles of equal area. According to Lemma 3.5.1, and the comment at the end of the proof, each of these 2^j triangles has area

$$\epsilon^2 \cdot 2^{-(j-1)} \left(\frac{\sqrt{3}/2}{j+1} \right)^2$$

$$\times \frac{1}{2\left[2\frac{\sqrt{3}}{2}\left(1 + \frac{1}{2} + \cdots + \frac{1}{j+1}\right) - \frac{\sqrt{3}}{2}\left(1 + \frac{1}{2} + \cdots + \frac{1}{j}\right)\right]}.$$

After k sproutings, the region we will have created will have area

$$\left[\sqrt{3}/4 + \sum_{j=1}^{k} 2^j \cdot 2^{-(j-1)} \right.$$

$$\left. \times \frac{[\sqrt{3}/2]/(j+1)^2}{2\left[2(1 + \frac{1}{2} + \cdots + \frac{1}{j+1}) - (1 + \frac{1}{2} + \cdots + \frac{1}{j})\right]} \right] \epsilon^2$$

$$\leq \left[\frac{\sqrt{3}}{4} + \sum_{j=1}^{k} \frac{\sqrt{3}}{2(j+1)^2} \right] \epsilon^2$$

$$\leq \left[\frac{\sqrt{3}}{2} \left(\frac{1}{2} + \left(\frac{\pi^2}{6} - 1 \right) \right) \right] \epsilon^2$$

$$\leq 2\epsilon^2.$$

We see that the set E has area as small as we please, just so long as the triangle with which we begin has small area. Further, the angles of the sprouted triangles span a total angle of $\pi/3$. Notice (Figure 10) that a segment of length two can be moved—incrementally—through this angle of $\pi/3$ by utilizing each sprouted triangle in succession. By adjoining a total of six of these sprout regions E—each rotated through an angle of $\pi/3$ from the previous one—we obtain a region inside which

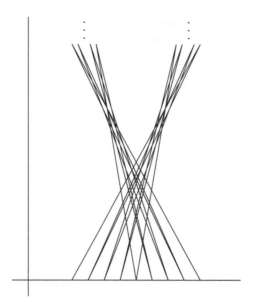

Figure 10.   Movement of a segment through the sprouts.

our needle (segment) can be rotated through a full turn. [Precisely, the
six sprouted triangles are joined so that all six of the initial triangles
from which they grew now have the same center; this "star" now becomes the "nucleus" of the six collections of sprouts. Place a needle
at the edge of one of the sprouts. The needle can be moved through a
small angle within that sprout, and then slid through the nucleus to a
sprout on the opposite side. Then the needle can be moved through a
small angle in this new sprout, and then slid through the nucleus to yet
another new sprout on the opposite side. And so forth.] And the area of
the set we have created does not exceed $6 \cdot 2\epsilon^2$. Since $\epsilon > 0$ is arbitrarily small, the Kakeya needle problem is solved. Refer to [CUN] for the
history of this fascinating problem and for all the details. That source
also describes many variants of the problem. The source [GUZ] has an
alternative, and very elegant, presentation of the Kakeya problem.

    While the solution of the Kakeya needle problem has intrinsic in-
terest, our purpose here is to obtain information about spherical sum-

mation of double trigonometric series. We shall now give Fefferman's arguments that establish the connection, and prove that spherical summation for multiple trigonometric series fails when $p \neq 2$. In the interest of brevity, and of not becoming bogged down in technical analytic arguments, we shall occasionally sketch proofs or refer the reader elsewhere for necessary details.

We begin with two functional analysis lemmas. They are elementary, but powerful. The first proves a "vector-valued" bound for an operator created from a single bounded linear operator.

Lemma 3.5.2 (A. Zygmund) *If S is a bounded linear operator on $L^p(\mathbb{R}^2)$, $1 \leq p < \infty$, then for functions $g_j \in L^p(\mathbb{R}^2)$, $j = 1, 2, \ldots, k$, we have*

$$\left\| \left(\sum_{j=1}^{k} |Sg_j|^2 \right)^{1/2} \right\|_{L^p} \leq C \left\| \left(\sum_{j=1}^{k} |g_j|^2 \right)^{1/2} \right\|_{L^p}.$$

The constant C is independent of the number k of g_j's, and depends only on the operator norm of S.

Proof. By linearity, we may assume that the g_j and Sg_j are real-valued. We assume that $g_1, \ldots, g_k \in L^p(\mathbb{R}^2)$ and we let $\alpha = (\alpha_1, \ldots, \alpha_k) \in \mathbb{R}^k$ satisfy $\sum \alpha_j^2 = 1$. Then, with C denoting the operator norm of S,

$$\int \left| \sum_j \alpha_j Sg_j \right|^p dx \leq C^p \int \left| \sum_j \alpha_j g_j \right|^p dx.$$

We may rewrite this inequality as

$$\int |(Sg_1(x), \ldots, Sg_k(x))|^p |\cos \theta_\alpha(x)|^p \, dx$$

$$\leq C^p \int |(g_1(x), \ldots, g_k(x))|^p |\cos \psi_\alpha(x)|^p \, dx.$$

Here $\theta_\alpha(x)$ is the angle between the vector $(Sg_1(x), \ldots Sg_k(x))$ and the vector $(\alpha_1, \ldots, \alpha_k)$ in $\mathbb{R}^k$ and $\psi_\alpha(x)$ is the angle between the vector $(g_1(x), \ldots g_k(x))$ and the vector $(\alpha_1, \ldots, \alpha_k)$ in $\mathbb{R}^k$.

Now we integrate both sides of the last inequality in α ranging over

$$\Sigma_k \equiv \left\{ \alpha \in \mathbb{R}^k : \sum_j \alpha_j^2 = 1 \right\},$$

using rotationally-invariant area measure (Appendix IV) on the unit sphere in $\mathbb{R}^k$. We apply Fubini's theorem to perform the α integration first and obtain

$$\int |(Sg_1(x), \ldots, Sg_k(x))|^p \, dx \le C^p \cdot \int |(g_1(x), \ldots, g_k(x))|^p \, dx.$$

Note in conclusion that the constant C in the last inequality is precisely equal to the operator norm of the original operator S. □

The lemma contains the inequality we seek for a finite collection of functions $g_1, \ldots, g_k$. We obtain the full result, for infinitely many g_j's, by passing to the limit.

Lemma 3.5.3 (Y. Meyer) *Let $\mathbf{v}_1, \mathbf{v}_2, \ldots$ be a sequence of unit vectors in $\mathbb{R}^2$. Let $H_j \equiv \{x \in \mathbb{R}^2 : x \cdot \mathbf{v}_j \ge 0\}$. We define operators $T_1, T_2, \ldots$ by*

$$\widehat{(T_j f)} \equiv \chi_{H_j} \cdot \widehat{f}.$$

Assume that the ball operator $T = \mathcal{M}_{\chi_B}$ defined by

$$\widehat{(T f)} \equiv \chi_B \cdot \widehat{f}$$

is bounded on L^p, some $1 < p < \infty$. Then, for any sequence of functions $f_j \in L^p(\mathbb{R}^2)$, we have

$$\left\| \left(\sum_j |T_j f_j|^2 \right)^{1/2} \right\|_p \le C \cdot \left\| \left(\sum_j |f_j|^2 \right)^{1/2} \right\|_p.$$

Proof. Let $D(j, r) = \{x \in \mathbb{R}^2 : |x - r\mathbf{v}_j| \le r\}$. We set $\mathcal{T}_{j,r} \equiv \mathcal{M}_{\chi_{D(j,r)}}$. If $f \in C_c^\infty(\mathbb{R}^2)$, then we see that, for every x,

$$
\begin{aligned}
\lim_{r \to \infty} \mathcal{T}_{j,r} f(x) &= \lim_{r \to \infty} \frac{1}{(2\pi)^2} \int_{\mathbb{R}^2} \widehat{f}(\xi) \chi_{D(j,r)}(\xi) e^{-i\xi \cdot x} \, d\xi \\
&= \frac{1}{(2\pi)^2} \int_{\mathbb{R}^2} \widehat{f}(\xi) \chi_{H_j}(\xi) e^{-i\xi \cdot x} \, d\xi \\
&= \mathcal{T}_j f(x).
\end{aligned}
$$

On the other hand, we may apply Fatou's lemma to obtain

$$
\left\| \left(\sum_{j=1}^k |\mathcal{T}_j f_j|^2 \right)^{1/2} \right\|_p \le \liminf_{r \to \infty} \left\| \left(\sum_{j=1}^k |\mathcal{T}_{j,r} f_j|^2 \right)^{1/2} \right\|_p
$$

as long as each $f_j \in C_c^\infty(\mathbb{R}^2)$. In order to obtain the desired estimate for finite sums and for $f_j \in C_c^\infty$ it thus suffices to see that

$$
\left\| \left(\sum_j |\mathcal{T}_{j,r} f_j|^2 \right)^{1/2} \right\|_p \le C \cdot \left\| \left(\sum_j |f_j|^2 \right)^{1/2} \right\|_p \tag{3.5.3.1}
$$

with C not depending on r.

In fact the independence from r follows from exploiting the dilation structure (i.e., a change of variable—see the proof of Lemma 3.4.1). So we may as well suppose that $r = 1$. Set $E_j(x) = e^{i\mathbf{v}_j \cdot x}$. We have

$$
\mathcal{T}_{j,1} f = \overline{E_j} \cdot \mathcal{T}(E_j \cdot f),
$$

so the left-hand side of (3.5.3.1) becomes (with $r = 1$)

$$
\left\| \left(\sum_{j=1}^k |\mathcal{T}(E_j \cdot f_j)|^2 \right)^{1/2} \right\|_p .
$$

By Zygmund's lemma, this last expression does not exceed

$$C \cdot \left\| \left(\sum_{j=1}^{k} |E_j \cdot f_j|^2 \right)^{1/2} \right\|_p = C \cdot \left\| \left(\sum_{j=1}^{k} |f_j|^2 \right)^{1/2} \right\|_p.$$

This proves the required result for $f_j \in C_c^\infty$ and for finite sums. The general case follows by a standard limiting argument. $\qquad\Box$

Here is how the remainder of our argument will play out: We shall do a tedious, but elementary, calculation of the size of the multiplier operator for a half-space operating on the characteristic function of a certain rectangle. We shall then piece a large number of these operations together, using the lemma of Meyer that we have just established. Finally, a contradiction will arise. What will we have contradicted? We will have contradicted the hypothesis of Meyer's lemma to the effect that the ball multiplier is bounded on some L^p, $p \neq 2$.

Lemma 3.5.4 *Let $R \subseteq \mathbb{R}^2$ be a rectangle, and let $\widetilde{R}$ denote the union of the two copies of R that are adjacent to R as shown in Figure 11. Assume that the long side of $R \cup \widetilde{R}$ (the side which contains parts of the boundaries of all three rectangles) runs parallel to the x_2-axis and the other side parallel to the x_1-axis. Assume further that R is centered at the origin.*

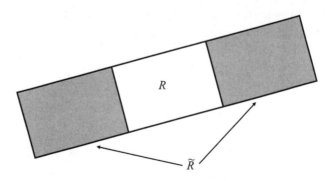

Figure 11. The region $\widetilde{R}$.

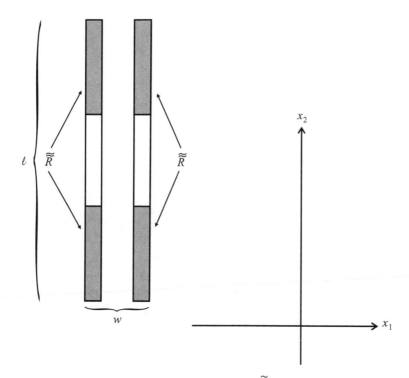

Figure 12. The region $\widetilde{\widetilde{R}}$.

Let ℓ be the length of the side of R that points in the x_2-direction and w the length of the other side (in the x_1-direction).

Set $\widetilde{\widetilde{R}} = \{x = (x_1, x_2) : x \in \widetilde{R}, w/4 < |x_1| < w/2\}$. See Figure 12.

Let $\mathbf{v}$ be a unit vector parallel to the longer side of $R \cup \widetilde{R}$, let $f = \chi_R$, and let $H = \{x \in \mathbb{R}^2 : x \cdot \mathbf{v} \geq 0\}$.

Set $S = \mathcal{M}_{\chi_H}$. Then $|Sf(x)| \geq 1/20$ for $x \in \widetilde{\widetilde{R}}$.

Proof. The truth of the lemma is independent of the shape of R; that this is so will become apparent from the proof. With this independence in mind, we shall assume for convenience that R is a *square* of side 2 with sides parallel to the axes and center 0, and that $\mathbf{v} = (0, 1)$.

We see that

$$\widehat{Sf}(\xi) = \chi_{\{\xi_2 \geq 0\}}(\xi) \int_{-1}^{1} e^{i\xi_1 t_1}\, dt_1 \int_{-1}^{1} e^{i\xi_2 t_2}\, dt_2$$

$$= \chi_{\{\xi_2 \geq 0\}}(\xi) \cdot 4 \cdot \frac{\sin \xi_1}{\xi_1} \cdot \frac{\sin \xi_2}{\xi_2}.$$

Restricting attention to $-1 < x_1 < 1$ and $|x_2| > 1$, we see that

$$Sf(x) = 4 \cdot (2\pi)^{-2} \int_0^\infty \int_{-\infty}^\infty \frac{\sin \xi_1}{\xi_1} \frac{\sin \xi_2}{\xi_2} e^{-i\xi \cdot x}\, d\xi_1 d\xi_2.$$

[Of course this last formula, and some of those below, will have to be taken to be formal. Gaussian summation, or some similar summation process, is required to make sure that the integrals converge.] Now to estimate $|Sf(x)|$ from below it suffices to estimate the real part of $Sf(x)$. Thus

$$\operatorname{Re} Sf(x) = \pi^{-2} \int_0^\infty \int_{-\infty}^\infty \frac{\sin \xi_1}{\xi_1} \frac{\sin \xi_2}{\xi_2} \cdot \cos(\xi \cdot x)\, d\xi_1 d\xi_2$$

$$= 2\pi^{-2} \int_0^\infty \int_0^\infty \frac{\sin \xi_1}{\xi_1} \frac{\sin \xi_2}{\xi_2} \cdot \cos(\xi \cdot x)\, d\xi_1 d\xi_2$$

$$= 2\pi^{-2} \int_0^\infty \int_0^\infty \frac{\sin \xi_1}{\xi_1} \frac{\sin \xi_2}{\xi_2} \cdot \left[\cos \xi_1 x_1 \cos \xi_2 x_2 \right.$$

$$\left. - \sin \xi_1 x_1 \sin \xi_2 x_2 \right] d\xi_1 d\xi_2$$

$$= \left[2\pi^{-2} \int_0^\infty \int_0^\infty \frac{[\sin \xi_1(1 + x_1) + \sin \xi_1(1 - x_1)]}{2\xi_1} \right.$$

$$\times \left. \frac{[\sin \xi_2(1 + x_2) + \sin \xi_2(1 - x_2)]}{2\xi_2}\, d\xi_1 d\xi_2 \right]$$

$$+ \left[2\pi^{-2} \int_0^\infty \int_0^\infty \frac{[\cos \xi_1(1 + x_1) - \cos \xi_1(1 - x_1)]}{2\xi_1} \right.$$

$$\times \frac{[\cos \xi_2(1 + x_2) - \cos \xi_2(1 - x_2)]}{2\xi_2} \, d\xi_1 d\xi_2 \Bigg]$$

$$\equiv A(x) + B(x).$$

Observe that

$$A(x) = 2\pi^{-2} \int_0^\infty \frac{[\sin \xi_1(1 + x_1) + \sin \xi_1(1 - x_1)]}{2\xi_1} \, d\xi_1$$

$$\times \int_0^\infty \frac{[\sin \xi_2(1 + x_2) + \sin \xi_2(1 - x_2)]}{2\xi_2} \, d\xi_2 ;$$

the second of these integrals is equal to 0 by a change of variable if $|x_2| > 1$. That disposes of the term $A(x)$.   □

Now, for $B(x)$, we have (remembering that $|x_1| < 1$)

$$\int_0^\infty \frac{\cos \xi_1(1 + x_1) - \cos \xi_1(1 - x_1)}{2\xi_1} \, d\xi_1$$

$$= \lim_{\substack{\epsilon \to 0 \\ \eta \to \infty}} \left[\int_{\epsilon(1+x_1)}^{\eta(1+x_1)} \frac{\cos \xi_1}{2\xi_1} \, d\xi_1 - \int_{\epsilon(1-x_1)}^{\eta(1-x_1)} \frac{\cos \xi_1}{2\xi_1} \, d\xi_1 \right]$$

$$= \lim_{\substack{\epsilon \to 0 \\ \eta \to \infty}} \left[\int_{\eta(1-x_1)}^{\eta(1+x_1)} \frac{\cos \xi_1}{2\xi_1} \, d\xi_1 - \int_{\epsilon(1-x_1)}^{\epsilon(1+x_1)} \frac{\cos \xi_1}{2\xi_1} \, d\xi_1 \right]$$

It is easy to see—integrating by parts—that the first of these integrals tends to zero as $\eta \to \infty$. The second integral is approximately equal to

$$\int_{\epsilon(1-x_1)}^{\epsilon(1+x_1)} \frac{1}{2\xi_1} \, d\xi_1,$$

and this integral tends to

$$\frac{1}{2} \log \left(\frac{1 + x_1}{1 - x_1} \right), \quad -1 < x_1 < 1.$$

That takes care of the first integral in $B(x)$.

The second integral in $B(x)$ is evaluated similarly, so that

$$|B(x)| = c \cdot \left| \log \left| \frac{1 + x_1}{1 - x_1} \right| \right| \left| \log \left| \frac{1 + x_2}{1 - x_2} \right| \right|.$$

Given our normalization of coordinates, the just-derived conclusion concerns points $x = (x_1, x_2)$ satisfying $1/2 < |x_1| < 1$ and $1 < |x_2| < 3$. For such x, we see that

$$|Sf(x)| \geq |\operatorname{Re} Sf(x)| = |B| \geq (\log 3)(\log 2)/10 \geq 1/20. \qquad \square$$

Lemma 3.5.5 *Let $\eta > 0$ be a small number. Then there is a set $E \subseteq \mathbb{R}^2$ and a collection $\mathcal{R} = \{R_j\}$ of pairwise disjoint rectangles such that*

(3.5.5.1) $|\widetilde{\widetilde{R}}_j \cap E| \geq |\widetilde{R}_j|/80;$

(3.5.5.2) $|E| \leq \eta \sum_j |R_j|.$

Here $|\ \ |$ represents Lebesgue 2-dimensional measure.

Proof. We continue the notation that was used in the solution of the Kakeya needle problem. Recall that we constructed a set E with area not exceeding $2\epsilon^2$; the construction was based on repeated sprouting of an initial equilateral triangle of side ϵ. For convenience, we now take $\epsilon = 1$.

Fix $k \in \{1, 2, \ldots\}$. Recall that each dyadic subinterval $I_j \equiv I_j^k \equiv [j/2^k, (j+1)/2^k] \subseteq [0, 1]$ is the base of precisely one sprouted triangle T_α, where $\alpha = (\alpha_0, \ldots, \alpha_k)$ has the property that $\alpha_0 = 0$ and each $\alpha_1, \ldots, \alpha_k$ is either 1 or 2. Call this triangle T_j. Let P_j be its upper vertex. Construct R_j as in Figure 13. Notice that R_j lies *below* the base of T_j and is uniquely determined by T_j and the fact that the side that is parallel to a side of T_j has length $2 \log k$. Now for each j, $\widetilde{\widetilde{R}}_j \cap E \supseteq \widetilde{\widetilde{R}}_j \cap T_j \equiv W_j$. Hence

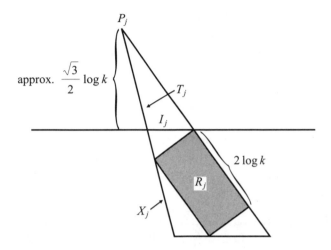

approx. $\dfrac{\sqrt{3}}{2}\log k$

P_j

T_j

I_j

$2\log k$

R_j

X_j

Figure 13. The rectangles R_j.

$$|\widetilde{\widetilde{R}}_j \cap E| \geq |W_j|$$

$$\geq \frac{1}{5} \cdot \frac{\sqrt{3}}{2} \cdot \log k \cdot 2^{-k}$$

$$\geq \frac{1}{80} \cdot 2 \cdot [6\log k \cdot 2^{-k}]$$

$$\geq \frac{|\widetilde{R}_j|}{80}.$$

This is condition (3.5.5.1) of the lemma.

On the other hand, each R_j has area not less than $[1/2]\cdot(\log k)2^{-k}$ and there are 2^k of them so (assuming that the R_j's are disjoint),

$$\left|\bigcup_{j=0}^{2^k-1} R_j\right| \geq \frac{1}{2}\log k \geq \frac{2}{\eta} \geq \frac{|E|}{\eta},$$

provided that we select $k > \exp(4/\eta)$. This is condition (3.5.5.2) of the lemma.

Finally, we check that $R_i \cap R_j = \emptyset$ if $i \neq j$. To see this, let us denote by X_j the trapezoid in Figure 13, which contains R_j. At the first sprouting, X_1^1 and X_2^1 are disjoint hence so are R_1 and R_2. Inductively, assume that the trapezoids $X_1^{k-1}, \ldots, X_{2^{k-1}}^{k-1}$ arising at the $(k-1)^{\text{th}}$ sprouting are disjoint. Consider X_i^k and X_j^k, with $i \neq j$, from the kth sprouting. In this circumstance either

(3.5.5.3) X_i^k and X_j^k are, respectively, subsets of different trapezoids from the $(k-1)^{\text{st}}$ sprouting, in which case they are *a fortiori* disjoint;

or else

(3.5.5.4) X_i^k and X_j^k are subsets of the same trapezoid from the $(k-1)^{\text{st}}$ sprouting. In this last case, they are disjoint by analogy with the first sprouting.

This discussion completes the proof of Lemma 3.5.5.   □

It remains to show that Lemma 3.5.5 completes the proof of Fefferman's theorem, that the ball multiplier cannot be bounded on $L^p(\mathbb{R}^2)$, for $p \neq 2$. By duality we may take $p > 2$ (see Sections 0.2, 0.3). Let E and $\mathcal{R} = \{R_j\}$ be as in Lemma 3.5.5. We let $f_j = \chi_{R_j}$. Let $\mathbf{v}_j$ be the unit vector parallel to the long side of $(R_j \cup \tilde{R}_j)$, each j. Let $H_j = \{x \in \mathbb{R}^2 : x \cdot \mathbf{v}_j \geq 0\}$ and $\mathcal{T}_j = \mathcal{M}_{\chi_{H_j}}$. Then (using Lemma 3.5.4)

$$\int_E \left(\sum_j |\mathcal{T}_j f_j(x)|^2 \right) dx = \sum_j \int_E |\mathcal{T}_j f_j(x)|^2 \, dx$$

$$\geq \frac{1}{400} \sum_j |E \cap \tilde{\tilde{R}}_j|$$

$$\geq \frac{1}{32000} \sum_j |\tilde{R}_j|$$

$$= \frac{1}{16000} \sum_j |R_j|. \qquad (3.5.5.5)$$

Applying Hölder's inequality to the functions χ_E and $(\sum_j |T_j f_j|^2)$ with exponents $p/(p-2)$ and $p/2$, we find that

$$\int_E \left(\sum_j |T_j f_j(x)|^2\right) dx \le |E|^{(p-2)/p} \left\|\left(\sum_j |T_j f_j|^2\right)^{1/2}\right\|_p^2 .$$

But if the conclusion of Meyer's lemma were true, then we would have that the last expression is

$$\le C|E|^{(p-2)/p} \left\|\left(\sum_j |f_j|^2\right)^{1/2}\right\|_p^2$$

$$= C|E|^{(p-2)/p} \left(\sum_j |R_j|\right)^{2/p}$$

$$\le C\eta^{(p-2)/p} \sum_j |R_j|. \tag{3.5.5.6}$$

Combining inequalities (3.5.5.5) and (3.5.5.6), we see that

$$\sum_j |R_j| \le 16000 C\eta^{(p-2)/p} \sum_j |R_j|.$$

This is absurd if η is small enough. That is the desired contradiction.

Let us summarize the rather elaborate arguments presented thus far in a theorem:

Theorem 3.5.6 *Let T be the multiplier operator for the ball:*

$$T : f \longmapsto (\chi_B \cdot \widehat{f})^{\vee},$$

Then T is bounded on L^2, but if $1 < p < \infty$ and $p \ne 2$, then T does not map $L^p(\mathbb{R}^N)$ into itself.

The reader will want to review the logic in the proof of this theorem—just the statements of the lemmas and theorems and propositions—to see how this contradiction came about. And notice that it can*not* come about if we are considering polygonal summation—

since dilates of a fixed polygon can only exhaust half planes that point in finitely many different directions. What goes wrong with the spherical summation operator is that the tangent lines to the disc (ball) point in infinitely many directions.

In a series of papers in the late 1970's, A. Cordoba and R. Fefferman endeavored to understand the phenomenon of non-convergence in the spherical sense. They proved that, for certain polygons with infinitely many sides, L^p convergence is still valid (see [CORF1], [CORF2], as well as [COR1], [COR2]). We are still a long way from understanding spherical convergence, and the related ideas of Bochner-Riesz summability (see [DAY]).

CHAPTER 4

Spherical Harmonics

4.1 A New Look at Fourier Analysis in the Plane

We learned in Section 1.1 that each character $e^{ik\theta}$ for the circle group has a unique harmonic extension to the disc. In fact, if $k \geq 0$, then the extension is the function z^k, and if $k < 0$, then the extension is the function $\bar{z}^{|k|}$. Thus we may view the Riesz-Fischer theorem as saying that any L^2 function on the circle may be expressed as a linear combination of boundary functions of the harmonic monomials $\{z^k\}_{k=0}^{\infty}$ and $\{\bar{z}^k\}_{k=1}^{\infty}$.

It is also natural to wish to decompose a function defined on all of $\mathbb{R}^2$ in terms of the characters $\{e^{ik\theta}\}_{k=-\infty}^{\infty}$. For specificity, let us concentrate on $L^2(\mathbb{R}^2)$. If $f \in L^2$ and if ρ is a rotation of the plane, then we define $(\rho f)(x) = f(\rho x)$. Thus the rotation group acts on the space of L^2 functions in an obvious way. We also know from Chapter 2 that the Fourier transform commutes with rotations: $\widehat{(\rho f)} = \rho \widehat{f}$.

An L^2 function f on $\mathbb{R}^2$ (or, more generally, on $\mathbb{R}^N$) is termed *radial* if $f(x) = f(y)$ whenever $|x| = |y|$. Another way of saying this is that f is *invariant under rotations*, or that $f(re^{i\theta})$ does not depend on θ (or that $\rho f = f$ for all ρ). We see from the calculations in the preceding paragraph that the Fourier transform preserves radial functions;

because of Plancherel's theorem, which says that the Fourier transform is an isometry on L^2, it is natural to suppose that the Fourier transform preserves the space (in L^2) that is orthogonal to radial functions. In fact we shall now obtain an orthogonal decomposition of $L^2(\mathbb{R}^2)$ into subspaces, each of which is preserved by the rotation group. In order to avoid some nasty measure-theoretic details, we shall treat the following calculation as purely formal. We invite the interested reader to fill in the details, or to refer to [STG1]. Note that, at various junctures in this chapter, we will find it convenient to identify a point (x, y) in the plane with the complex number $z = x + iy$ and then in turn with the polar complex number $re^{i\theta}$, where $r = \sqrt{x^2 + y^2}$ and θ is some argument of z. We do so without further comment.

Fix a function $f \in L^2(\mathbb{R}^2)$. For each fixed $r > 0$, we may consider the function $f_r(e^{i\theta}) = f(re^{i\theta})$, which (by Fubini's theorem) is in L^2 of the circle for almost every $r \geq 0$. Thus, by Riesz-Fischer theory, we may write

$$f(re^{i\theta}) = f_r(e^{i\theta}) = \sum_{j=-\infty}^{\infty} f_{r,j} e^{ij\theta}.$$

Notice that $f_{r,j}$ is the jth Fourier coefficient for the function f_r. By Parseval,

$$\sum_{j=-\infty}^{\infty} |f_{j,r}|^2 = \frac{1}{2\pi} \int_0^{2\pi} |f_r(e^{i\theta})|^2 \, d\theta = \frac{1}{2\pi} \int_0^{2\pi} |f(re^{i\theta})|^2 \, d\theta.$$

By Lebesgue's monotone convergence theorem,

$$\lim_{M \to \infty} \int_0^\infty \sum_{-M}^{M} |f_{j,r}|^2 r \, dr = \frac{1}{2\pi} \int_0^\infty \left\{ \int_0^{2\pi} |f(re^{i\theta})|^2 \, d\theta \right\} r \, dr$$

$$= \frac{1}{2\pi} \|f\|_{L^2}^2.$$

As a result, if we set $g_j(re^{i\theta}) = f_{r,j} e^{ij\theta}$ for $j = 0, \pm 1, \pm 2, \ldots, z = re^{i\theta}$, then we have

$$\lim_{M \to \infty} \int_{\mathbb{R}^2} \left| f(z) - \sum_{j=-M}^{M} g_j(z) \right|^2 dx dy$$

$$= \lim_{M \to \infty} 2\pi \int_0^\infty \left[\frac{1}{2\pi} \left\{ \int_0^{2\pi} |f(re^{i\theta})|^2 d\theta \right\} - \sum_{j=-M}^{M} |f_{j,r}|^2 \right] r \, dr$$

$$= 0.$$

Observing that the characters $e^{ij\theta}$ and $e^{ik\theta}$ are orthogonal when $j \neq k$, we conclude that L^2 has the direct sum decomposition

$$L^2(\mathbb{R}^2) = \bigoplus_{k=-\infty}^{\infty} \mathcal{H}_k,$$

where

$$\mathcal{H}_k \equiv \{g \in L^2(\mathbb{R}^2) : g(re^{i\theta}) = f(r)e^{ik\theta},$$

$$\text{some function } f \text{ satisfying } \int_0^\infty |f(r)|^2 r \, dr < \infty\}.$$

Observe that this notation is consistent with our earlier observations: for $\mathcal{H}_0$ is just the space of radial, square-integrable functions. Further, the orthogonal complement of $\mathcal{H}_0$ is the direct sum of the spaces $\mathcal{H}_k$, $k \neq 0$—each of which is mapped to itself by the Fourier transform.

To confirm this last statement, we let f be a testing function in $\mathcal{H}_k$. So $f(re^{i\theta}) = e^{ik\theta} \Phi(r)$. Fix an angle ϕ and set $g(x) = f(e^{i\phi}x)$. It follows, with $x = re^{i\theta}$, that $g(x) = e^{ik\phi} f(x)$. But recall that the Fourier transform commutes with rotations. It follows that

$$\widehat{f}(e^{i\phi}\xi) = \widehat{\rho_\phi f}(\xi) = \widehat{g}(\xi) = e^{ik\phi} \widehat{f}(\xi).$$

Now we take $\xi = r > 0$. We find that

$$\widehat{f}(re^{i\phi}) = e^{ik\phi} \widehat{f}(r).$$

Thus $\widehat{f} \in \mathcal{H}_k$, as we wished to prove.

In $\mathbb{R}^N$, when seeking a generalization of planar Fourier analysis, we may take the approach of looking at the restrictions to the unit sphere of harmonic polynomials on all of space. These spherical harmonics are for many purposes the natural generalization of the Fourier analysis of the circle to higher dimensions. Spherical harmonics are also intimately connected to the representation theory of the orthogonal group. As a result, analogues of the spherical harmonics play an important role in general representation theory.

In this discussion we shall use *multi-index* notation. Here, on N-dimensional Euclidean space, α is a multi-index if $\alpha = (\alpha_1, \ldots, \alpha_N)$ is an N-tuple of nonnegative integers. For $x = (x_1, x_2, \ldots, x_N)$ we define

$$x^\alpha \equiv x_1^{\alpha_1} x_2^{\alpha_2} \cdots x_N^{\alpha_N}.$$

Also

$$\frac{\partial^\alpha}{\partial x^\alpha} \equiv \frac{\partial}{\partial x_1^{\alpha_1}} \frac{\partial}{\partial x_2^{\alpha_2}} \cdots \frac{\partial}{\partial x_N^{\alpha_N}}.$$

It is common to use $|\alpha|$ to denote[1] $\alpha_1 + \cdots \alpha_N$ and to let $\alpha!$ denote $\alpha_1! \alpha_2! \cdots \alpha_N!$. We use the notation $\delta_{\alpha\beta}$ to denote the *Kronecker delta* for multi-indices: $\delta_{\alpha\beta} = 1$ if $\alpha = \beta$ and $\delta_{\alpha\beta} = 0$ if $\alpha \neq \beta$.

Our presentation of spherical harmonics owes a debt to [STG1]. For $k = 0, 1, 2, \ldots$ we let $\mathcal{P}_k$ denote the linear space over $\mathbb{C}$ of all homogeneous polynomials in $\mathbb{R}^N$ of degree k. Then $\{x^\alpha\}_{|\alpha|=k}$ is a basis for $\mathcal{P}_k$. Let d_k denote the dimension, over the field $\mathbb{C}$, of $\mathcal{P}_k$. We need to calculate d_k. This will require a counting argument.

Fix the dimension N of $\mathbb{R}^N$. We need to determine the number of N-tuples $\alpha = (\alpha_1, \ldots, \alpha_N)$ such that $\alpha_1 + \cdots \alpha_N = k$. Imagine $N + k - 1$ boxes as shown in Figure 1. We shade any $N - 1$ of these boxes. Let $\alpha_1 \geq 0$ be the number of white boxes preceding the first one shaded, $\alpha_2 \geq 0$ be the number of white boxes between the first and second that are shaded, and so on. This defines N nonnegative integers $\alpha_1, \ldots, \alpha_N$ such that $\alpha_1 + \cdots + \alpha_N = k$. Also every such N-tuple

---

[1] In mathematical analysis, vertical bars are used in a number of different ways. The use to measure multi-indices is different from the use to measure points in Euclidean space. Context will make clear which meaning is intended.

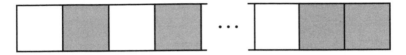

Figure 1.   Counting multi-indices.

$(\alpha_1, \ldots, \alpha_N)$ arises in this way. Thus we see that

$$d_k = \binom{N+k-1}{N-1} = \binom{N+k-1}{k} = \frac{(N+k-1)!}{(N-1)!k!}.$$

Now we want to define a Hermitian inner product on $\mathcal{P}_k$. In this section, if $P(x) = \sum_\alpha c_\alpha x^\alpha$ is a polynomial, then the differential operator $P(D)$ is defined to be

$$P(D) = \sum_\alpha c_\alpha \frac{\partial^\alpha}{\partial x^\alpha}.$$

For $P, Q \in \mathcal{P}_k$ we then define

$$\langle P, Q \rangle \equiv P(D)\big[\,\overline{Q}\,\big].$$

If $P(x) = \sum_{|\alpha|=k} p_\alpha x^\alpha$ and $Q(x) = \sum_{|\alpha|=k} q_\alpha x^\alpha$, then we have

$$\begin{aligned}
\langle P, Q \rangle = P(D)\big[\,\overline{Q}\,\big] &= \sum_{|\alpha|=k} p_\alpha \partial^\alpha \left(\sum_{|\beta|=k} \overline{q}_\beta x^\beta \right) \\
&= \sum_{|\alpha|,|\beta|=k} p_\alpha \overline{q}_\beta \partial^\alpha x^\beta \\
&= \sum_{|\alpha|,|\beta|=k} p_\alpha \overline{q}_\beta \delta_{\alpha\beta} \alpha!
\end{aligned}$$

Therefore $\langle P, Q \rangle$ is scalar-valued. It is (real) linear in each entry and Hermitian symmetric. Moreover, we see that

$$\langle P, P \rangle = \sum_\alpha |p_\alpha|^2 \alpha!$$

so that

$$\langle P, P \rangle \geq 0 \quad \text{for all } P \quad \text{and} \quad \langle P, P \rangle = 0 \quad \text{iff} \quad P = 0.$$

Thus $\langle \, \cdot \, , \, \cdot \, \rangle$ is a Hermitian, nondegenerate (i.e., positive definite) inner product on $\mathcal{P}_k$.

Proposition 4.1.1 *Let $P \in \mathcal{P}_k$. Then we can write*

$$P(x) = P_0(x) + |x|^2 P_1(x) + \cdots + |x|^{2\ell} P_\ell(x),$$

where each polynomial P_j is homogeneous and harmonic with degree $k - 2j, 0 \leq j \leq \ell$, and $\ell = [k/2]$.

Proof. Any polynomial of degree less than 2 is harmonic, so there is nothing to prove in this case. We therefore assume that $k \geq 2$. Define the map

$$\phi_k : \mathcal{P}_k \to \mathcal{P}_{k-2}$$

$$P \mapsto \triangle P,$$

where $\triangle$ is the (classical) Laplacian,

$$\triangle \phi \equiv \sum_{j=1}^{N} \frac{\partial^2}{\partial x_j^2} \phi.$$

Now consider the adjoint operator (calculated with respect to the inner product just defined)

$$\phi_k^* : \mathcal{P}_{k-2} \to \mathcal{P}_k.$$

This adjoint is determined by the equalities

$$\langle Q, \triangle P \rangle = Q(D)\left[\overline{\triangle P} \right] = \triangle Q(D)\overline{P} = \langle R, P \rangle,$$

where $R(x) = |x|^2 Q(x)$. Therefore

$$\phi_k^*(Q)(x) = R(x) = |x|^2 Q(x).$$

Notice that ϕ_k^* is one-to-one. The identity

$$\langle Q, \phi_k(P) \rangle = \langle \phi_k^*(Q), P \rangle \quad \text{for all } Q \in \mathcal{P}_{k-2}, P \in \mathcal{P}_k$$

shows that $\ker \phi_k$ and $\operatorname{im} \phi_k^*$ are orthogonal complements in the space $\mathcal{P}_k$:

$$\mathcal{P}_k = \ker \phi_k \oplus \operatorname{im} \phi_k^*.$$

That is,

$$\mathcal{P}_k = \mathcal{A}_k \oplus \mathcal{B}_k,$$

where

$$\mathcal{A}_k = \ker \phi_k = \{ P \in \mathcal{P}_k : \triangle P = 0 \}$$

and

$$\mathcal{B}_k = \operatorname{im} \phi_k^* = \{ P \in \mathcal{P}_k : P(x) = |x|^2 Q(x), \text{ some } Q \in \mathcal{P}_{k-2} \}.$$

Hence, for $P \in \mathcal{P}_k$,

$$P(x) = P_0(x) + |x|^2 Q(x)$$

where P_0 is harmonic and $Q \in \mathcal{P}_{k-2}$.

The result now follows immediately by induction. $\qquad\square$

Corollary 4.1.2 *The restriction to the surface of the unit sphere Σ_{N-1} of any polynomial of N variables is a sum of restrictions to Σ_{N-1} of harmonic polynomials.*

Proof. Use the preceding proposition. The expressions $|x|^{2j}$ become 1 when restricted to the sphere. $\qquad\square$

Definition 4.1.3 The *spherical harmonics* of degree k, denoted $\mathcal{H}_k$, are the restrictions to the unit sphere of the elements of $\mathcal{A}_k$, the homogeneous harmonic polynomials of degree k.

The space $\mathcal{A}_k$ is called the space of *solid spherical harmonics* and the space $\mathcal{H}_k$ is the space of *surface spherical harmonics*.

If $Y = P\big|_{\Sigma_{N-1}}$ for some $P \in \mathcal{A}_k$, then

$$P(x) = Y(x/|x|) \cdot |x|^k$$

so that the restriction is an isomorphism of $\mathcal{A}_k$ onto $\mathcal{H}_k$. In particular,

$$\begin{aligned}
\dim \mathcal{H}_k &= \dim \mathcal{A}_k \\
&= \dim \mathcal{P}_k - \dim \mathcal{P}_{k-2} \\
&= d_k - d_{k-2} \\
&= \binom{N+k-1}{k} - \binom{N+k-3}{k-2}
\end{aligned}$$

for $k \geq 2$. Notice that $\dim \mathcal{H}_0 = 1$ and $\dim \mathcal{H}_1 = N$.

For $N = 2$, it is easy to see that

$$\mathcal{H}_k = \text{span}\left\{\cos k\theta, \sin k\theta\right\}.$$

Then $\dim \mathcal{H}_k = 2$ for all $k \geq 1$. This is of course consistent with the formula for the dimension of $\mathcal{H}_k$ that we just derived for all dimensions. For $N = 3$, one sees that $\dim \mathcal{H}_k = 2k + 1$ for all $k \geq 0$. We denote $\dim \mathcal{H}_k = \dim \mathcal{A}_k$ by the symbol a_k. [It is common to let σ denote surface measure on the boundary of a domain; in the present instance, σ is rotationally invariant surface measure on the unit sphere.]

Proposition 4.1.4 *The finite linear combinations of elements of $\cup_k \mathcal{H}_k$ are uniformly dense in $C(\Sigma_{N-1})$ and L^2-dense in $L^2(\Sigma_{N-1}, d\sigma)$.*

Proof. The first statement clearly implies the second. For the first we invoke the Stone-Weierstrass theorem (Appendix VIII). To wit, the restriction of any polynomial to the sphere is a linear combination of restrictions of harmonic polynomials to the sphere. But the set of all polynomials clearly vanishes nowhere and separates points. So Stone-Weierstrass applies. $\qquad\square$

Proposition 4.1.5 *If $Y^{(k)} \in \mathcal{H}_k$ and $Y^{(\ell)} \in \mathcal{H}_\ell$ with $k \neq \ell$ then*

$$\int_{\Sigma_{N-1}} Y^{(k)}(x') Y^{(\ell)}(x') \, d\sigma(x') = 0.$$

Proof. We will use Green's theorem (see Appendix V or [KRA4]): If $u, v \in C^2(\overline{\Omega})$, where Ω is a bounded domain with C^2 boundary, then

$$\int_{\partial\Omega} \left(u \frac{\partial}{\partial \nu} v - v \frac{\partial}{\partial \nu} u \right) d\sigma = \int_{\Omega} (u \,\triangle\, v - v \,\triangle\, u) \, dV.$$

Here $\partial/\partial\nu$ is the (unit) outward normal derivative to $\partial\Omega$.

Now for $x \in \mathbb{R}^N$ we write $x = rx'$ with $r = |x|$ and $|x'| = 1$. Then

$$u(x) \equiv |x|^k Y^{(k)}(x')$$

and

$$v(x) \equiv |x|^\ell Y^{(\ell)}(x')$$

are harmonic polynomials.

If one of k or ℓ is zero, then one of u or v is constant and what we are about to do reduces to the well-known fact that, for a harmonic function f on B, C^1 on $\overline{B}$, we have

$$\int_{\partial B} \frac{\partial}{\partial \nu} f \, d\sigma = 0$$

(see [KRA4]). Details are left for the reader.

In case both k and ℓ are nonzero, then on Σ_{N-1} we have

$$\frac{\partial}{\partial \nu} u(x') = \frac{\partial}{\partial r} \left(r^k Y^{(k)}(x') \right)$$

$$= k r^{k-1} Y^{(k)}(x')$$

$$= k Y^{(k)}(x')$$

(since $r = 1$) and, similarly,

$$\frac{\partial}{\partial \nu} v(x') = \ell Y^{(\ell)}(x').$$

By Green's theorem (Appendix V) and the harmonicity of u and v,

$$
\begin{aligned}
0 &= \int_B \left[u(x) \, \triangle \, v(x) - v(x) \, \triangle \, u(x) \right] dV(x) \\
&= \int_{\partial B} \left[u \frac{\partial}{\partial \nu} v - v \frac{\partial}{\partial \nu} u \right] d\sigma \\
&= \int_{\partial B} \left[u(x') \ell Y^{(\ell)}(x') - v(x') k Y^{(k)}(x') \right] d\sigma(x') \\
&= \int_{\partial B} \left[\ell Y^{(k)}(x') Y^{(\ell)}(x') - k Y^{(\ell)}(x') Y^{(k)}(x') \right] d\sigma(x') \\
&= (\ell - k) \int_{\partial B} Y^{(k)}(x') Y^{(\ell)}(x') \, d\sigma(x').
\end{aligned}
$$

Since $\ell \neq k$, the assertion follows. $\square$

We endow $L^2(\partial B, d\sigma)$ with the usual inner product given by

$$\langle f, g \rangle = \int_{\Sigma_{N-1}} f(x) \overline{g(x)} \, d\sigma(x).$$

So of course each $\mathcal{H}_k$ inherits this inner product as well. For $k = 0, 1, 2, \ldots$, we let $\{ Y_1^{(k)}, \ldots, Y_{a_k}^{(k)} \}$, $a_k = d_k - d_{k-2}$, be an orthonormal basis for $\mathcal{H}_k$. By Propositions 4.1.4 and 4.1.5 it follows that

$$\bigcup_{k=0}^{\infty} \{ Y_1^{(k)}, \ldots, Y_{a_k}^{(k)} \}$$

is an orthonormal basis for $L^2(\Sigma_{N-1}, d\sigma)$. Each $f \in L^2(\Sigma_{N-1})$ has a unique representation

$$f = \sum_{k=0}^{\infty} Y^{(k)},$$

where the series converges in the L^2 topology and $Y^{(k)} \in \mathcal{H}_k$. Furthermore, by orthonormality,

$$Y^{(k)} = \sum_{j=1}^{a_k} b_j Y_j^{(k)},$$

where $b_j = \langle Y^{(k)}, Y_j^{(k)} \rangle$, $j = 1, \ldots, a_k$.

As an example of these ideas, we see for $N = 2$ that

$$k = 0 : \qquad Y_1^{(0)} \equiv \frac{1}{\sqrt{2\pi}}$$

$$k \geq 1 : \qquad \begin{cases} Y_1^{(k)}(\theta) = \frac{1}{\sqrt{\pi}} \cos k\theta \\ Y_2^{(k)}(\theta) = \frac{1}{\sqrt{\pi}} \sin k\theta. \end{cases}$$

Here, for convenience, we are using polar coordinates. Thus we have that

$$\left\{ \frac{1}{\sqrt{2\pi}}, \frac{1}{\sqrt{\pi}} \cos k\theta, \frac{1}{\sqrt{\pi}} \sin k\theta \right\}$$

is a complete orthonormal system in $L^2(\mathbb{T})$.

Claim: We can, in any dimension, recover the Poisson kernel for the Laplacian from the spherical harmonics. Let us perform the calculation in dimension 2. More will be said about the higher-dimensional situation in the next section.

For $f \in L^2(\partial D)$ and $0 \leq r < 1$ define

$$F(re^{i\theta}) = \sum_{j,k} r^k \langle f, Y_j^{(k)} \rangle Y_j^{(k)}(e^{i\theta}).$$

Then we have

$$F(re^{i\theta}) = \int_0^{2\pi} f(e^{i\phi}) \frac{1}{\sqrt{2\pi}} \, d\phi \cdot \frac{1}{\sqrt{2\pi}}$$

$$+ \sum_{k=1}^{\infty} r^k \int_0^{2\pi} f(e^{i\phi}) \frac{\cos k\phi}{\sqrt{\pi}} \, d\phi \frac{\cos k\theta}{\sqrt{\pi}}$$

$$+ \sum_{k=1}^{\infty} r^k \int_0^{2\pi} f(e^{i\phi}) \frac{\sin k\phi}{\sqrt{\pi}} \, d\phi \frac{\sin k\theta}{\sqrt{\pi}}$$

$$= \frac{1}{2\pi} \int_0^{2\pi} f(e^{i\phi}) \, d\phi$$

$$+ \frac{1}{\pi} \sum_{k=1}^{\infty} r^k \int_0^{2\pi} f(e^{i\phi}) \cos k(\theta - \phi) \, d\phi$$

$$= \frac{1}{\pi} \int_0^{2\pi} f(e^{i\phi}) \left[\frac{1}{2} + \sum_{k=1}^{\infty} r^k \cos k(\theta - \phi) \right] d\phi.$$

But the expression in brackets equals

$$\frac{1}{2} + \text{Re} \left\{ \sum_{k=1}^{\infty} r^k e^{ik(\theta - \phi)} \right\}$$

$$= \frac{1}{2} + \text{Re} \left\{ r e^{i(\theta - \phi)} \cdot \sum_{k=0}^{\infty} r^k e^{ik(\theta - \phi)} \right\}$$

$$= \frac{1}{2} + \text{Re} \left\{ r e^{i(\theta - \phi)} \cdot \frac{1}{1 - r e^{i(\theta - \phi)}} \right\}$$

$$= \frac{1}{2} \cdot \frac{1 - r^2}{1 - 2r \cos(\theta - \phi) + r^2}.$$

Thus

$$F(re^{i\theta}) = \frac{1}{2\pi} \int_0^{2\pi} \frac{1 - r^2}{1 - 2r \cos(\theta - \phi) + r^2} f(e^{i\phi}) \, d\phi.$$

It follows from elementary Hilbert space considerations (see also the discussion at the end of Section 1.4) that $F(re^{i\theta}) \to f(e^{i\theta})$ in L^2 of the circle as $r \to 1^-$. [First check this claim on finite linear combinations of spherical harmonics, which are dense; see also [GRK].] Thus, at least formally, we have recovered the classical Poisson integral formula from spherical harmonic analysis.

4.2   Further Results on Spherical Harmonics

In this section we cover some more advanced topics in the theory of spherical harmonics. Part of the purpose is to extend the results at the end of the preceding section from dimension 2 to dimension $N \geq 3$. Another motivation is to introduce the reader to some of the machinery of higher-dimensional harmonic analysis.

Indeed, many of the special functions of mathematical physics arise rather naturally in the context of multidimensional harmonic analysis. In this section we shall meet, for instance, the Gegenbauer polynomials and the Bessel functions.

Although the results of this section are not intrinsically difficult, they are highly calculational and rather technical. The reader may wish, for a first pass, to just consider the statements of results and to save mastering the proofs for a later reading. These results will *not* be used in later chapters of the book.

We assume from now on that $N > 2$. Fix a point $x' \in \Sigma_{N-1}$ and consider the linear functional on $\mathcal{H}_k$ given by

$$e_{x'} : Y \mapsto Y(x').$$

Of course $\mathcal{H}_k$ is a finite-dimensional Hilbert space (with the standard Hermitian inner product) so there exists a unique spherical harmonic $Z_{x'}^{(k)}$ such that

$$Y(x') = e_{x'}(Y) = \int_{\Sigma_{N-1}} Y(t') Z_{x'}^{(k)}(t') \, d\sigma(t')$$

for all $Y \in \mathcal{H}_k$.

Definition 4.2.1 The function $Z_{x'}^{(k)}$ is called the *zonal harmonic* of degree k with pole at x'.

Lemma 4.2.2 *If $\left\{ Y_1, \ldots, Y_{a_k} \right\}$ is an orthonormal basis for $\mathcal{H}_k$ then, for all $x', t' \in \Sigma_{N-1}$ and all rotations ρ,*

(4.2.2.1) $\sum_{m=1}^{a_k} \overline{Y_m(x')} Y_m(t') = Z_{x'}^{(k)}(t')$;

(4.2.2.2) $Z_{x'}^{(k)}$ is real-valued and $Z_{x'}^{(k)}(t') = Z_{t'}^{(k)}(x')$;

(4.2.2.3) $Z_{\rho x'}^{(k)}(\rho t') = Z_{x'}^{(k)}(t')$.

Proof. The standard representation of $Z_{x'}^{(k)}$ with respect to the orthonormal basis $\{Y_1, \ldots, Y_{a_k}\}$ is $Z_{x'}^{(k)} = \sum_{m=1}^{a_k} \langle Z_{x'}^{(k)}, Y_m \rangle Y_m$, where

$$\langle Z_{x'}^{(k)}, Y_m \rangle = \int_{\Sigma_{N-1}} \overline{Y_m(t')} Z_{x'}^{(k)}(t') \, d\sigma(t') = \overline{Y_m(x')}.$$

We have used here the reproducing property of the zonal harmonic (note that, since Y_m is harmonic, so is $\overline{Y_m}$). This proves (4.2.2.1), for we now know that

$$Z_{x'}^{(k)}(t') = \sum_{m=1}^{a_k} \langle Z_{x'}^{(k)}, Y_m \rangle Y_m(t') = \sum_{m=1}^{a_k} \overline{Y_m(x')} Y_m(t').$$

To prove (4.2.2.2), let $f \in \mathcal{H}_k$. Then

$$\overline{f}(x') = \int_{\Sigma_{N-1}} \overline{f}(t') Z_{x'}^{(k)}(t') \, d\sigma(t')$$

$$= \overline{\int_{\Sigma_{N-1}} f(t') \overline{Z_{x'}^{(k)}(t')} \, d\sigma(t')}.$$

That is,

$$f(x') = \int_{\Sigma_{N-1}} f(t') \overline{Z_{x'}^{(k)}(t')} \, d\sigma(t').$$

Thus we see that $\overline{Z_{x'}^{(k)}}$ reproduces $\mathcal{H}_k$ at the point x'. By the uniqueness of the zonal harmonic at x', we conclude that $Z_{x'}^{(k)} = \overline{Z_{x'}^{(k)}}$. That is, $Z_{x'}^{(k)}$ is real-valued. Now, using (4.2.2.1), we have

$$Z_{x'}^{(k)}(t') = \sum_{m=1}^{a_k} \overline{Y_m(x')} Y_m(t')$$

$$= \overline{\sum_{m=1}^{a_k} Y_m(x')\overline{Y_m(t')}}$$

$$= \overline{Z_{t'}^{(k)}(x')}$$

$$= Z_{t'}^{(k)}(x').$$

This establishes (4.2.2.2).

To check that (4.2.2.3) holds, it suffices by uniqueness to see that $Z_{\rho x'}^{(k)}(\rho t')$ reproduces $\mathcal{H}_k$ at x'. This is a formal exercise which we omit. [The matter comes down to the ρ-invariance of the surface measure σ and the closure of $\mathcal{H}_k$ under the operation $f \mapsto \rho f$.]   □

Lemma 4.2.3 *Let* $\{Y_1, \ldots, Y_{a_k}\}$ *be any orthonormal basis for* $\mathcal{H}_k$. *The following properties hold for the zonal harmonics, and for all* $x', t' \in \Sigma_{N-1}$:

(4.2.3.1) $Z_{x'}^{(k)}(x') = a_k/\sigma(\Sigma_{N-1})$, *where* $a_k = \dim \mathcal{A}_k = \dim \mathcal{H}_k$;

(4.2.3.2) $\sum_{m=1}^{a_k} |Y_m(x')|^2 = a_k/\sigma(\Sigma_{N-1})$;

(4.2.3.3) $|Z_{t'}^{(k)}(x')| \leq a_k/\sigma(\Sigma_{N-1})$;

(4.2.3.4) $\|Z_{x'}^{(k)}\|_{L^2}^2 = a_k/\sigma(\Sigma_{N-1})$.

Proof. Let $x_1', x_2' \in \Sigma_{N-1}$ and let ρ be a rotation such that $\rho x_1' = x_2'$. Then by parts (4.2.2.1) and (4.2.2.3) we know (think of a rotation that takes x_1 to x_2) that

$$\sum_{m=1}^{a_k} |Y_m(x_1')|^2 = Z_{x_1'}^{(k)}(x_1') = Z_{x_2'}^{(k)}(x_2') = \sum_{m=1}^{a_k} |Y_m(x_2')|^2 \equiv c.$$

Then

$$a_k = \sum_{m=1}^{a_k} \int_{\Sigma_{N-1}} |Y_m(x')|^2 \, d\sigma(x')$$

$$= \int_{\Sigma_{N-1}} \sum_{m=1}^{a_k} |Y_m(x')|^2 \, d\sigma(x')$$

$$= c\sigma(\Sigma_{N-1}).$$

This proves parts (4.2.3.1) and (4.2.3.2).

For part (4.2.3.4), notice that

$$\|Z_{x'}^{(k)}\|_{L^2}^2 = \int_{\Sigma_{N-1}} |Z_{x'}^{(k)}(t')|^2 \, d\sigma(t')$$

$$= \int_{\Sigma_{N-1}} \left(\sum_m \overline{Y_m(x')} Y_m(t') \right) \left(\overline{\sum_\ell \overline{Y_\ell(x')} Y_\ell(t')} \right) \, d\sigma(t')$$

$$= \sum_m |Y_m(x')|^2$$

$$= \frac{a_k}{\sigma(\Sigma_{N-1})}.$$

Finally, we use the reproducing property of the zonal harmonics together with (4.2.3.4) to see that

$$|Z_{t'}^{(k)}(x')| = \left| \int_{\Sigma_{N-1}} Z_{t'}^{(k)}(w') Z_{x'}^{(k)}(w') \, d\sigma(w') \right|$$

$$\leq \|Z_{t'}^{(k)}\|_{L^2} \cdot \|Z_{x'}^{(k)}\|_{L^2}$$

$$= \frac{a_k}{\sigma(\Sigma_{N-1})}. \qquad \Box$$

Now we wish to present a version of the expansion of the Poisson kernel in terms of spherical harmonics in higher dimensions (see Section 4.1 for the case of dimension 2). The Poisson kernel for the ball B in $\mathbb{R}^N$ is

$$P(x, t') = \frac{1}{\sigma(\Sigma_{N-1})} \frac{1 - |x|^2}{|x - t'|^N}$$

for $0 \leq |x| < 1$ and $|t'| = 1$ (see [KRA4] for an independent derivation of this formula). In the discussion below, we take it for granted that the

Poisson kernel, as given by this formula, solves the Dirichlet problem on the ball B. This means that, for each $f \in C(\partial B)$, the function

$$u(x) \equiv \int_{\partial B} P(x, t') f(t') \, d\sigma(t')$$

satisfies

(4.2.4) The function u is harmonic on B.

(4.2.5) The function u extends continuously to $\overline{B}$.

(4.2.6) When u is extended as in (4.2.5), the restriction of u to the boundary of B equals f.

We say that the function u, so defined, solves the Dirichlet problem on the ball B with data f.

Now we have

Theorem 4.2.7 *For $x \in B$ we write $x = rx'$ with $|x'| = 1$. Then*

$$P(x, t') = \sum_{k=0}^{\infty} r^k Z_{x'}^{(k)}(t') = \sum_{k=0}^{\infty} r^k Z_{t'}^{(k)}(x')$$

is the Poisson kernel for the ball.

Proof. Observe that, for N fixed,

$$a_k = d_k - d_{k-2} = \binom{N+k-1}{k} - \binom{N+k-3}{k-2}$$

$$= \frac{(N+k-3)!}{(k-1)!(N-2)!}$$

$$\times \left\{ \frac{(N+k-1)(N+k-2)}{k(N-1)} - \frac{k-1}{N-1} \right\}$$

$$= \binom{N+k-3}{k-1} \left\{ \frac{N+2k-2}{k} \right\}$$

$$\leq C_N \cdot \binom{N+k-3}{k-1}$$

$$\leq C_N \cdot k^{N-2}.$$

Here $C = C_N$ depends on the dimension N, but not on k. With this estimate, and the estimate on the size of the zonal harmonics from the preceding lemma, we see that the series

$$\sum_{k=0}^{\infty} r^k Z_{t'}^{(k)}(x')$$

converges uniformly on compact subsets of B. In fact we may establish the stronger assertion that the series converges uniformly on $\Sigma_{N-1} \times K$ for any compact subset $K \subseteq B$. Indeed, for $|x| \leq s < 1, x = rx'$, $t' \in \Sigma_{N-1}$, we have that

$$\sum_{k=0}^{\infty} |r^k Z_{t'}^{(k)}(x')| \leq \sum_{k=0}^{\infty} s^k \frac{a_k}{\sigma(\Sigma_{N-1})} \leq \sum_{k=0}^{\infty} s^k \frac{C \cdot k^{N-2}}{\sigma(\Sigma_{N-1})}$$

$$= C' \cdot \sum_{k=0}^{\infty} s^k k^{N-2} < \infty.$$

Now fix k. Let $X = \sum_{m=0}^{p} X_m$ be a finite linear combination of spherical harmonics with all $X_m \in \mathcal{H}_k$. Then

$$\sum_{m=0}^{p} |x|^k X_m \left(\frac{x}{|x|}\right) \equiv u(x) = \int_{\Sigma_{N-1}} X(t') P(x, t') \, d\sigma(t')$$

is the solution to the classical Dirichlet problem with data X. On the other hand

$$\int_{\Sigma_{N-1}} X(t') \sum_{k=0}^{\infty} r^k Z_{t'}^{(k)}(x') \, d\sigma(t')$$

$$= \sum_{m=0}^{p} \int_{\Sigma_{N-1}} X_m(t') \sum_{k=0}^{\infty} |x|^k Z_{t'}^{(k)}(x') \, d\sigma(t')$$

$$= \sum_{m=0}^{p} \sum_{k=0}^{\infty} |x|^k \int_{\Sigma_{N-1}} X_m(t') Z_{t'}^{(k)}(x') \, d\sigma(t')$$

$$= \sum_{m=0}^{p} |x|^m X_m(x')$$

$$= u(x).$$

Thus

$$\int_{\Sigma_{N-1}} \left[P(x, t') - \sum_{k} r^k Z_{t'}^{(k)}(x') \right] X(t') \, d\sigma(t') = 0$$

for all $x \in B$ and for all finite linear combinations X of spherical harmonics. Since k was arbitrary, the finite linear combinations of spherical harmonics of all degrees are dense in $L^2(\Sigma_{N-1})$. The desired assertion follows.   □

Our immediate goal now is to obtain an explicit formula for each zonal harmonic $Z_{x'}^{(k)}$. We begin this process with some generalities about polynomials.

Lemma 4.2.8 *Let P be a polynomial in $\mathbb{R}^N$ such that*

$$P(\rho x) = P(x)$$

for all $\rho \in O(N)$ (the orthogonal group) and $x \in \mathbb{R}^N$. Then there exist constants $c_0, \ldots, c_p$ such that

$$P(x) = \sum_{m=0}^{p} c_m \left(x_1^2 + \cdots + x_N^2 \right)^m.$$

Proof. We write P as a sum of homogeneous terms:

$$P(x) = \sum_{\ell=0}^{q} P_\ell(x),$$

where P_ℓ is homogeneous of degree ℓ. Now for any $\epsilon > 0$ and $\rho \in O(N)$ we have

$$\sum_{\ell=0}^{q} \epsilon^\ell P_\ell(x) = \sum_{\ell=0}^{q} P_\ell(\epsilon x)$$

$$= P(\epsilon x)$$

$$= P(\epsilon \rho x)$$

$$= \sum_{\ell=0}^{q} P_\ell(\epsilon \rho x)$$

$$= \sum_{\ell=0}^{q} \epsilon^\ell P_\ell(\rho x).$$

For fixed x, we think of the far left and far right of this last sequence of equalities as identities *of polynomials in* ϵ. It follows that $P_\ell(x) = P_\ell(\rho x)$ for every ℓ. The result of these calculations is that we may concentrate our attentions on P_ℓ.

Consider the function $|x|^{-\ell} P_\ell(x)$. It is homogeneous of degree 0 and still invariant under the action of $O(N)$. Then

$$|x|^{-\ell} P_\ell(x) = c_\ell,$$

for some constant c_ℓ. This forces ℓ to be even (since P_ℓ is a *polynomial* function); the result follows.   □

Definition 4.2.9 Let $\eta \in \Sigma_{N-1}$. A *parallel of* Σ_{N-1} *orthogonal to* η is the intersection of Σ_{N-1} with a hyperplane (not necessarily through the origin) orthogonal to the line determined by η and the origin.

Notice that a parallel of Σ_{N-1} orthogonal to η is a set of the form

$$\{x' \in \Sigma_{N-1} : x' \cdot \eta = c\},$$

$-1 \leq c \leq 1$. Observe that a function F on Σ_{N-1} is constant on parallels orthogonal to $\eta \in \Sigma_{N-1}$ if and only if for all $\rho \in O(N)$ that fix η and all $x' \in \Sigma_{N-1}$ the equation $F(\rho x') = F(x')$ holds.

Lemma 4.2.10 *Let* $\eta \in \Sigma_{N-1}$. *An element* $Y \in \mathcal{H}_k$ *is constant on parallels of* Σ_{N-1} *orthogonal to* η *if and only if there exists a constant c such that*

$$Y = cZ_\eta^{(k)}.$$

Proof. Recall that we are assuming that $N \geq 3$. Let ρ be a rotation that fixes η. Then, for each $x' \in \Sigma_{N-1}$, we have

$$Z_\eta^{(k)}(x') = Z_{\rho\eta}^{(k)}(\rho x') = Z_\eta^{(k)}(\rho x').$$

Hence $Z_\eta^{(k)}$ is constant on the parallels of Σ_{N-1} orthogonal to η.

To prove the converse direction, assume that $Y \in \mathcal{H}_k$ is constant on the parallels of Σ_{N-1} orthogonal to η. Let $e_1 = (1, 0, \ldots, 0) \in \Sigma_{N-1}$ and let τ be a rotation such that $\eta = \tau e_1$. Define

$$W(x') = Y(\tau x').$$

Then $W \in \mathcal{H}_k$ is constant on the parallels of Σ_{N-1} orthogonal to e_1. Suppose we can show that $W = cZ_{e_1}^{(k)}$ for some constant c. Then, for all $x' \in \Sigma_{N-1}$,

$$Y(x') = W(\tau^{-1}x') = cZ_{e_1}^{(k)}(\tau^{-1}x')$$

$$= cZ_{\tau e_1}^{(k)}(x') = cZ_\eta^{(k)}(x').$$

So the lemma will follow. Thus we examine W and take $\eta = e_1$.
Define

$$P(x) = \begin{cases} |x|^k W(x/|x|) & \text{if } x \neq 0 \\ 0 & \text{if } x = 0. \end{cases}$$

Let ρ be a rotation that fixes e_1. We write

$$P(x) = \sum_{j=0}^{k} x_1^{k-j} P_j(x_2, \ldots, x_N).$$

Since ρ fixes the powers of x_1, it follows that ρ leaves each P_j invariant. Then each P_j is a polynomial in $(x_2, \ldots, x_N) \in \mathbb{R}^{N-1}$ that is invariant under the rotations of $\mathbb{R}^{N-1}$. We conclude (see the last paragraph of the proof of Lemma 4.2.8) that $P_j = 0$ for odd j and

$$P_j(x_2, \ldots, x_N) = c_j \left(x_2^2 + \cdots + x_N^2 \right)^{j/2} \equiv c_j R^j (x_2, \ldots, x_N)$$

for j even, where $R(x_2, \ldots, x_N) = \sqrt{x_2^2 + \cdots + x_N^2}$. Therefore

$$P(x) = c_0 x_1^k + c_2 x_1^{k-2} R^2 + \cdots c_{2\ell} x_1^{k-2\ell} R^{2\ell},$$

for some $\ell \leq k/2$. Of course P is harmonic, so $\triangle P \equiv 0$. A direct calculation then shows that

$$0 = \triangle P = \sum_p \left[c_{2p} \alpha_p + c_{2(p+1)} \beta_p \right] x_1^{k-2(p+1)} R^{2p},$$

where

$$\alpha_p \equiv (k - 2p)(k - 2p - 1)$$

and

$$\beta_p \equiv 2(p + 1)(N + 2p - 1).$$

Therefore we find the following recursion relation for the c's:

$$c_{2(p+1)} = -\frac{\alpha_p c_{2p}}{\beta_p} \qquad \text{for } p = 0, 1, \ldots, \ell - 1.$$

In particular, c_0 determines all the other c's.

From this it follows that all the elements of $\mathcal{H}_k$ that are constant on parallels of Σ_{N-1} orthogonal to e_1 are constant multiples of each other. Since $Z_{e_1}^{(k)}$ is one such element of $\mathcal{H}_k$, this proves our result.

$\square$

Lemma 4.2.11 *Fix k. Let $F_{y'}(x')$ be defined for all $x', y' \in \Sigma_{N-1}$. Assume that*

(4.2.11.1) *The function $F_{y'}(\,\cdot\,)$ is a spherical harmonic of degree k for every $y' \in \Sigma_{N-1}$;*

(4.2.11.2) *For every rotation ρ we have $F_{\rho y'}(\rho x') = F_{y'}(x')$, all $x', y' \in \Sigma_{N-1}$.*

Then there is a constant c such that, for all $x', y' \in \Sigma_{N-1}$,

$$F_{y'}(x') = c Z_{y'}^{(k)}(x').$$

Proof of the Lemma. Fix $y' \in \Sigma_{N-1}$ and let $\rho \in O(N)$ be such that $\rho(y') = y'$. Then

$$F_{y'}(x') = F_{\rho y'}(\rho x') = F_{y'}(\rho x').$$

Therefore, by the preceding lemma,

$$F_{y'}(x') = c_{y'} Z_{y'}^{(k)}(x') \qquad \text{for all } x' \in \Sigma_{N-1}.$$

(Here the constant $c_{y'}$ may in principle depend on y'). We need to see that, for $y'_1, y'_2 \in \Sigma_{N-1}$ arbitrary, $c_{y'_1} = c_{y'_2}$. Let $\sigma \in O(N)$ be such that $\sigma(y'_1) = y'_2$. By hypothesis (4.2.11.2),

$$
\begin{aligned}
c_{y'_2} Z_{y'_2}^{(k)}(\sigma x') &= F_{y'_2}(\sigma x') \\
&= F_{\sigma y'_1}(\sigma x') \\
&= F_{y'_1}(x') \\
&= c_{y'_1} Z_{y'_1}^{(k)}(x') \\
&= c_{y'_1} Z_{\sigma y'_1}^{(k)}(\sigma x') \\
&= c_{y'_1} Z_{y'_2}^{(k)}(\sigma x').
\end{aligned}
$$

Since these equalities hold for all $x' \in \Sigma_{N-1}$, we conclude that

$$c_{y'_2} = c_{y'_1}.$$

That is, calling this common constant c,

$$F_{y'}(x') = c Z_{y'}^{(k)}(x').$$ □

Definition 4.2.12 Let $0 \leq |z| < 1, |t| \leq 1$, and fix $\lambda > 0$. Consider the equation $z^2 - 2tz + 1 = 0$. Then $z = t \pm \sqrt{t^2 - 1}$ so that $|z| = 1$. Hence $z^2 - 2tz + 1$ is zero-free in the disc $\{z : |z| < 1\}$, and the function $z \mapsto (1 - 2tz + z^2)^{-\lambda}$ is well-defined and holomorphic in the disc. Set, for $0 \leq r < 1$,

$$(1 - 2rt + r^2)^{-\lambda} = \sum_{k=0}^{\infty} P_k^{\lambda}(t) r^k.$$

Then $P_k^{\lambda}(t)$ is said to be the *Gegenbauer polynomial of degree k* associated to the parameter λ.

Proposition 4.2.13 *The Gegenbauer polynomials enjoy the following properties:*

(4.2.13.1) $P_0^{\lambda}(t) \equiv 1$;

(4.2.13.2) $\frac{d}{dt} P_k^{\lambda}(t) = 2\lambda P_{k-1}^{\lambda+1}(t)$ *for* $k \geq 1$;

(4.2.13.3) $\frac{d}{dt} P_1^{\lambda}(t) = 2\lambda P_0^{\lambda+1}(t) = 2\lambda$;

(4.2.13.4) P_k^{λ} *is actually a polynomial of degree k in t;*

(4.2.13.5) *The monomials $1, t, t^2, \ldots$ can be obtained as finite linear combinations of $P_0^{\lambda}, P_1^{\lambda}, P_2^{\lambda}, \ldots$;*

(4.2.13.6) *The linear space spanned by the P_k^{λ}'s is uniformly dense in $C[-1, 1]$;*

(4.2.13.7) $P_k^{\lambda}(-t) = (-1)^k P_k^{\lambda}(t)$ *for all $k \geq 0$.*

Proof. We obtain (4.2.13.1) by simply setting $r = 0$ in the defining equation for the Gegenbauer polynomials.

For (4.2.13.2), note that

$$2r\lambda \sum_{k=0}^{\infty} P_k^{\lambda+1}(t) r^k \equiv 2r\lambda (1 - 2rt + r^2)^{-(\lambda+1)}$$

$$= \frac{d}{dt}\left(1 - 2rt + r^2\right)^{-\lambda}$$

$$= \sum_{k=0}^{\infty} \frac{d}{dt} P_k^\lambda(t) r^k.$$

The result now follows by identifying coefficients of like powers of r.
For (4.2.13.3), observe that (using (4.2.13.1) and (4.2.13.2))

$$\frac{d}{dt} P_1^\lambda(t) = 2\lambda P_0^{\lambda+1}(t) = 2\lambda.$$

It follows from integration that P_1^λ is a polynomial of degree 1 in t. Applying (4.2.13.2) and iterating yields (4.2.13.4).

Now (4.2.13.5) follows from (4.2.13.4) (inductively), and (4.2.13.6) is immediate from (4.2.13.5) and the Weierstrass approximation theorem.

Finally,

$$\sum_{k=0}^{\infty} P_k^\lambda(-t) r^k \equiv \left(1 - 2r(-t) + r^2\right)^{-\lambda}$$

$$= \left(1 - 2t(-r) + (-r)^2\right)^{-\lambda}$$

$$= \sum_{k=0}^{\infty} P_k^\lambda(t)(-r)^k$$

$$= \sum_{k=0}^{\infty} (-1)^k P_k^\lambda(t) r^k.$$

Now (4.2.13.7) follows from comparing coefficients of like powers of r.

$\square$

Theorem 4.2.14 *Let* $N > 2$, $\lambda = (N-2)/2$, $k \in \{0, 1, 2, \ldots\}$. *Then there exists a constant* $c_{k,N}$ *such that, for all* x', $y' \in \Sigma_{N-1}$,

$$Z_{y'}^{(k)}(x') = c_{k,N} P_k^\lambda(x' \cdot y').$$

Exercise. Compute by hand what the analogous formula is for $N = 2$. (Recall that the zonal harmonics in dimension 2 are just $\cos k\theta / \sqrt{\pi}$ and $\sin k\theta / \sqrt{\pi}$ for $k \geq 1$.)

Proof. We will use the formula for the Poisson integral on the ball (right before Theorem 4.2.7) to derive the result from Theorem 4.2.7. I thank G. B. Folland for this argument.

We apply the differential operator $(r/\lambda)(d/dr) + 1$ to the formula

$$\sum_{k=0}^{\infty} P_k^{\lambda}(t) r^k = (1 - 2rt + r^2)^{-\lambda}$$

to obtain

$$\sum \left(\frac{k}{\lambda} + 1 \right) r^k P_k^{\lambda}(t) = \frac{1 - r^2}{(1 - 2rt + r^2)^{\lambda+1}}. \qquad (4.2.14.1)$$

On the other hand, for all $x', y' \in \Sigma_{N-1}$,

$$P(rx', y') = \frac{1}{\sigma(\Sigma_{N-1})} \frac{1 - r^2}{(1 - 2rx' \cdot y' + r^2)^{N/2}}.$$

Using equation (4.2.14.1) with $t = x' \cdot y'$ and $\lambda = (N - 2)/2$, we find that

$$\sum_k r^k Z_{y'}^{(k)}(x') = P(rx', y')$$

$$= \sum_k \frac{2k + N - 2}{(N - 2)\sigma(\Sigma_{N-1})} r^k P_k^{(N-2)/2}(x' \cdot y').$$

Comparing coefficients of r^k yields the desired formula:

$$Z_{y'}^{(k)}(x') = \frac{2k + N - 2}{(N - 2)\sigma(\Sigma_{N-1})} P_k^{(N-2)/2}(x' \cdot y'). \qquad \square$$

We have seen how to express the Poisson kernel for the ball—in any dimension—in terms of the zonal harmonics. In turn, we have expressed the zonal harmonics in terms of the Gegenbauer polynomials.

We conclude this section by recording an important formula for the Fourier transform of a radial function. In this way we return to the theme by means of which we first introduced spherical harmonics.

We know that a function f on $\mathbb{R}^N$ is radial if $f(\rho x) = f(x)$ for any rotation ρ of space. If f is also in L^1, then we may consider the action of rotations on $\widehat{f}$. We find that

$$\rho \widehat{f} = \widehat{\rho f} = \widehat{f};$$

hence $\widehat{f}$ is radial. It turns out that an analogous assertion is true for each of the spaces $\mathcal{H}_k$. In fact the Fourier transform maps each $\mathcal{H}_k$ to itself in a natural fashion. Thus the decomposition of L^2 into the direct sum of the spaces $\mathcal{H}_k$ is a decomposition into invariant subspaces. We now formulate this statement in some detail and discuss the proof.

First we must define the Bessel functions. For each integer k we define the kth *Bessel function* by the formula

$$J_k(t) = \frac{1}{2\pi} \int_0^{2\pi} e^{it \sin \theta} e^{-ik\theta} \, d\theta \,, \qquad \text{for all } t \in \mathbb{R}.$$

More generally, for any real number k exceeding $-1/2$, we set

$$J_k(t) = \frac{(t/2)^k}{\Gamma[(2k+1)/2]\Gamma(\frac{1}{2})} \int_{-1}^1 e^{its}(1 - s^2)^{(2k-1)/2} \, ds.$$

Of course Γ is the standard special function of classical analysis ($\Gamma(r) \equiv \int_0^\infty e^{-t} t^{r-1} \, dt$). We refer the reader to [GRK, Ch. 15] for details. A proof of the equivalence of these two definitions of the Bessel function may be found in [WAT].

Now we have

Proposition 4.2.15 *Let* $f \in L^2(\mathbb{R}^N) \cap L^1(\mathbb{R}^N)$ *have the form* $f(x) = \phi(|x|) \cdot P(x)$, *where* ϕ *is a function of one real variable and* P *is a solid spherical harmonic of degree* k. *Then* $\widehat{f}$ *has the form* $\widehat{f}(\xi) = \Phi(|\xi|) \cdot P(\xi)$, *where the function* Φ *of one real variable is given by*

$$\Phi(r) = 2\pi i^{-k} r^{(N+2k-2)/2} \int_0^\infty \phi(s) J_{(N+2k-2)/2}(2\pi rs) s^{(N+2k)/2} \, ds.$$

In particular, we see that each space of solid spherical harmonics is mapped to itself by the Fourier transform.

Proof. The proof for general N involves tedious calculations that would take us far afield. Therefore we content ourselves with a look at dimension 2.

Let $f(x) = \phi(r)e^{ik\theta}$, where $x = re^{i\theta}$. Of course $F \equiv \widehat{f}$ has a similar form (see the beginning of Section 4.1), so that $F(\xi) = \Phi(R)e^{ik\psi}$, where $\xi = Re^{i\psi}$. Let us now endeavor to calculate Φ:

We have

$$\Phi(R) = F(Re^{i0})$$

$$= \int_0^\infty \phi(r) \left\{ \int_0^{2\pi} e^{iRr\cos\theta} e^{ik\theta} \, d\theta \right\} r \, dr.$$

In the first exponential, we have noted that

$$(R, 0) \cdot (r\cos\theta, r\sin\theta) = Rr\cos\theta.$$

With the change of variable $\theta \mapsto \theta - \pi/2$, this last becomes

$$(-i)^k 2\pi \int_0^\infty \phi(r) \left\{ \frac{1}{2\pi} \int_0^{2\pi} e^{iRr\sin\theta} e^{ik\theta} \, d\theta \right\} r \, dr.$$

Of course the expression in braces is the Bessel function J_{-k}. Thus, in summary,

$$\Phi(R) = (-i)^k 2\pi \int_0^\infty \phi(r) J_{-k}(rR) r \, dr.$$

This completes the calculation.

We leave it to the reader to compare the formula we have actually derived with the one enunciated in the statement of the proposition. Establishing that they are really equivalent (for $k > 0$) will involve certain difficult calculations with the Bessel functions—see [STG1, p. 158, Theorem 3.10] and also [WAT]. $\qquad\square$

CHAPTER **5**

Fractional Integrals, Singular Integrals, and Hardy Spaces

5.1 Fractional Integrals and Other Elementary Operators

For $\phi \in C_c^1(\mathbb{R}^N)$ we know that

$$\widehat{\frac{\partial \phi}{\partial x_j}}(\xi) = -i\xi_j \cdot \widehat{\phi}(\xi). \tag{5.1.1}$$

In other words, the Fourier transform converts differentiation in the x-variable to multiplication by a monomial in the Fourier transform variable. Of course higher-order derivatives correspond to multiplication by higher-order monomials.

It is natural to wonder whether the Fourier transform can provide us with a way to think about differentiation to a fractional order. In pursuit of this goal, we begin by thinking about the Laplacian

$$\triangle \phi \equiv \sum_{j=1}^{N} \frac{\partial^2}{\partial x_j^2} \phi.$$

Of course formula (5.1.1) shows that

$$\widehat{\triangle \phi}(\xi) = -|\xi|^2 \widehat{\phi}(\xi). \tag{5.1.2}$$

199

In the remainder of this section, let us use the notation

$$\mathcal{D}^2 \phi(\xi) = -\triangle \phi(\xi).$$

Then we set $\mathcal{D}^4 \phi \equiv \mathcal{D}^2 \circ \mathcal{D}^2 \phi$, and so forth.

Now let us examine the Fourier transform of $\mathcal{D}^2 \phi$ from a slightly more abstract point of view. Observe that the operator $\mathcal{D}^2$ is translation-invariant. Therefore, by the Schwartz kernel theorem [SCH], it is given by a convolution kernel k_2. Thus

$$\mathcal{D}^2 \phi(x) = \phi * k_2(x).$$

Therefore

$$\widehat{\mathcal{D}^2 \phi}(\xi) = \widehat{\phi}(\xi) \cdot \widehat{k_2}(\xi).$$

If we wish to understand differentiation from the point of view of the Fourier transform, then we should calculate $\widehat{k_2}$ and then k_2.

Notice that $\mathcal{D}^2$ commutes with rotations. It follows that k_2 is rotationally invariant, and hence so is $\widehat{k_2}$ (exercise). Also the operator $\mathcal{D}^2$ is homogeneous of degree $-N - 2$ (refer to the notion of homogeneity in the weak sense in Chapter 2), hence so is the kernel k_2. By Proposition 2.2.9, we can be sure that $\widehat{k_2}$ is homogeneous of degree 2. It follows that

$$\widehat{k_2}(\xi) = c \cdot |\xi|^2$$

for some constant c, consistent with our calculations in (5.1.2). [Since the expression on the right neither vanishes at infinity—think of the Riemann-Lebesgue lemma—nor is it in L^2, the reader will have to treat the present calculations as purely formal.] In other words,

$$\widehat{\mathcal{D}^2 \phi}(\xi) = c \cdot |\xi|^2 \cdot \widehat{\phi}(\xi).$$

More generally,

$$\widehat{\mathcal{D}^{2j} \phi}(\xi) = c^j \cdot |\xi|^{2j} \cdot \widehat{\phi}(\xi).$$

The calculations presented thus far should be considered to have been a finger exercise. Making them rigorous would require a considerable dose of the theory of Schwartz distributions, and this we wish to avoid. Now we enter the more rigorous phase of our discussion.

It turns out to be more efficient to study fractional integration than fractional differentiation. This is only a technical distinction, but the kernels that arise in the theory of fractional integration are a bit easier to study. Thus, in analogy with the operators $\mathcal{D}^2$, we define $\mathcal{I}^2\phi$ according to the identity

$$\widehat{\mathcal{I}^2\phi}(\xi) = |\xi|^{-2} \cdot \widehat{\phi}(\xi).$$

Observing that this Fourier multiplier is rotationally invariant and homogeneous of degree -2, we conclude that the *kernel* corresponding to the fractional integral operator $\mathcal{I}^2$ is $k_2(x) = |x|^{-N+2}$. [For convenience, we suppress any constant that may belong in front of this kernel.] Thus

$$\mathcal{I}^2\phi(x) = \int_{\mathbb{R}^N} |t|^{-N+2}\phi(x - t)\, dt$$

—at least when $N > 2$.

More generally, if $0 < \beta < N$, we *define*

$$\mathcal{I}^\beta\phi(x) = \int |t|^{-N+\beta}\phi(x - t)\, dt$$

for any testing function $\phi \in C_c^1(\mathbb{R}^N)$. Observe that this integral is absolutely convergent—near the origin because ϕ is bounded and $R_\beta(x) = |x|^{-N+\beta}$ is integrable, and near ∞ because ϕ is compactly supported. The operators $\mathcal{I}^\beta$ are called *fractional integral operators*.

Now the basic fact about fractional integration is that it acts naturally on the L^p spaces, for p in a particular range. Indeed we may anticipate exactly what the correct theorem is by using a little dimensional analysis. Fix $0 < \beta < N$ and suppose that an inequality of the form

$$\|\mathcal{I}^\beta\phi\|_{L^q} \le C \cdot \|\phi\|_{L^p}$$

were true for all testing functions ϕ. Let us replace ϕ in both sides of this inequality by the expression $\alpha_\delta \phi(x) \equiv \phi(\delta x)$. Writing out the integrals, we have

$$\left(\int_{\mathbb{R}^N} \left| \int_{\mathbb{R}^N} |t|^{\beta - N} \alpha_\delta \phi(x - t)\, dt \right|^q dx \right)^{1/q}$$

$$\leq C \cdot \left(\int_{\mathbb{R}^N} |\alpha_\delta \phi(x)|^p\, dx \right)^{1/p}.$$

On the left side we replace t by t/δ and x by x/δ; on the right we replace x by x/δ. The result, after a little calculation, is

$$\left(\delta^{-\beta q - N} \right)^{1/q} \left(\int_{\mathbb{R}^N} \left| \int_{\mathbb{R}^N} |t|^{\beta - N} \phi(x - t)\, dt \right|^q dx \right)^{1/q}$$

$$\leq \delta^{-N/p} C \cdot \left(\int_{\mathbb{R}^N} |\phi(x)|^p\, dx \right)^{1/p}.$$

Since this inequality must hold for any fixed ϕ and for all $\delta > 0$, there would be a contradiction either as $\delta \to 0$ or as $\delta \to +\infty$ if the two expressions in δ were not equal.

We conclude that

$$\left(\delta^{-\beta q - N} \right)^{1/q} = \delta^{-N/p}.$$

After a little algebra with the exponents, we find that

$$\frac{1}{q} = \frac{1}{p} - \frac{\beta}{N}.$$

Thus our calculation points to the correct theorem:

Theorem 5.1.3 *Let $0 < \beta < N$. The integral operator*

$$\mathcal{I}^\beta \phi(x) \equiv \int_{\mathbb{R}^N} |t|^{-N + \beta} \phi(x - t)\, dt,$$

initially defined for $\phi \in C_c^1(\mathbb{R}^N)$, satisfies

$$\|\mathcal{I}^\beta \phi\|_{L^q(\mathbb{R}^N)} \leq C \cdot \|\phi\|_{L^p(\mathbb{R}^N)},$$

whenever $1 < p < N/\beta$ *and* q *satisfies*

$$\frac{1}{q} = \frac{1}{p} - \frac{\beta}{N}.$$

We shall provide a proof of (a very general version of) this result in Theorem 6.2.1.

Fractional integrals are one of the two building blocks of the theory of integral operators that has developed in the last half century. In the next section we introduce the other building block.

5.2   Prolegomena to Singular Integral Theory

The Hilbert transform (Sections 1.1, 1.2) is the quintessential example of a singular integral. [We shall treat singular integrals in detail in Section 6.2 ff.] In fact in dimension 1 it is, up to multiplication by a constant, the only classical singular integral. This statement means that the function $1/t$ is the only integral kernel that (**i**) is smooth away from 0, (**ii**) is homogeneous of degree -1, and (**iii**) has "mean value 0" on the unit sphere in $\mathbb{R}^1$.

In $\mathbb{R}^N$, $N > 1$, there are a great many singular integral operators. Let us give a formal definition (refer to [CALZ]):

Definition 5.2.1   A function $K : \mathbb{R}^N \rightarrow \mathbb{C}$ is called a *Calderón-Zygmund singular integral kernel* if it possesses the following three properties:

(**5.2.1.1**)   The function K is smooth on $\mathbb{R}^N \setminus \{0\}$;

(**5.2.1.2**)   The function K is homogeneous of degree $-N$;

(**5.2.1.3**) $\int_{\Sigma_{N-1}} K(x)\, d\sigma(x) = 0$, where Σ_{N-1} is the $(N-1)$-dimensional unit sphere in $\mathbb{R}^N$, and $d\sigma$ is rotationally invariant surface measure on that sphere.

It is worthwhile to put this definition in context. Let β be a fixed complex number and consider the functional

$$\phi \longmapsto \int \phi(x)|x|^\beta \, dx,$$

which is defined on functions ϕ that are C^∞ with compact support. When $\operatorname{Re} \beta > -N$, this definition makes good sense, because we may estimate the integral near the origin (taking $\operatorname{supp}(\phi) \subseteq B(0, R)$, $C = \sup |\phi|$, and $C' = C \cdot \sigma(\Sigma_{N-1})$) by

$$C \cdot \left| \int_{\{|x| \leq R\}} |x|^\beta \, dx \right| \leq C \cdot \int_{\{|x| \leq R\}} |x|^{\operatorname{Re} \beta} \, dx$$

$$= C' \cdot \int_0^R r^{\operatorname{Re} \beta + N - 1} \, dr < \infty.$$

Now we change our point of view; we think of the test function ϕ as being fixed and we think of $\beta \in \mathbb{C}$ as the variable. In fact Morera's theorem shows that

$$\mathcal{G}(\beta) \equiv \int \phi(x)|x|^\beta \, dx$$

is well-defined and is a holomorphic function of β on $\{\beta \in \mathbb{C} : \operatorname{Re} \beta > -N\}$. We may ask whether this holomorphic function can be analytically continued to the rest of the complex plane.

In order to carry out the needed calculation, it is convenient to assume that the test function ϕ is a radial function: $\phi(x) = \phi(x')$ whenever $|x| = |x'|$. In this case we may write $\phi(x) = f(r)$, where $r = |x|$. Then we may write, using polar coordinates,

$$\mathcal{G}(\beta) \equiv \int \phi(x)|x|^\beta \, dx = c \cdot \int_0^\infty f(r)r^\beta \cdot r^{N-1} \, dr.$$

Integrating by parts in r gives

$$\mathcal{G}(\beta) = -\frac{c}{\beta + N} \int_0^\infty f'(r)r^{\beta+N} \, dr.$$

Notice that the boundary term at infinity vanishes since ϕ (and hence f) is compactly supported; the boundary term at the origin vanishes because of the presence of $r^{\beta+N}$.

We may continue, in this fashion, integrating by parts to obtain the formulas

$$
\mathcal{G}(\beta) = \frac{(-1)^{j+1}}{(\beta + N)(\beta + N + 1) \cdots (\beta + N + j)}
$$
$$
\times \int_0^\infty f^{(j+1)}(r) r^{\beta+N+j} \, dr. \qquad (5.2.2_j)
$$

The key fact is that any two of these formulas for $\mathcal{G}(\beta)$ are equal for $\operatorname{Re} \beta > -N$. Yet the integral in formula $(5.2.2_j)$ makes sense for $\operatorname{Re} \beta > -N - j - 1$. Hence formula $(5.2.2_j)$ can be used to *define* an extension of $\mathcal{G}$ to the domain $\operatorname{Re} \beta > -N - j - 1$. As a result, we have a method of analytically continuing $\mathcal{G}$ to the entire complex plane, less the poles at $\{-N, -N - 1, -N - 2, \ldots\}$. Observe that these poles are exhibited explicitly in the denominator of the fraction preceding the integral in the formulas $(5.2.2_j)$ that define $\mathcal{G}$.

The upshot of our calculations is that it is possible to make sense of the operator consisting of integration against $|x|^\beta$ as a classical fractional integral operator provided that $\beta \neq -N, -N - 1, \ldots$. More generally, an operator with integral kernel homogeneous of degree β, where $\beta \neq -N, -N - 1, \ldots$, is amenable to a relatively simple analysis.

If instead we consider an operator with kernel homogeneous of degree β where β takes on one of these critical values $-N, -N - 1, \ldots$, then some additional condition must be imposed on the kernel. These observations give rise to the mean-value-zero condition (5.2.1.3) in the definition of the Calderón-Zygmund integrals that are homogeneous of degree $-N$. [The study of singular integrals of degree $-N-k, k > 0$, is an advanced topic (known as the theory of strongly singular integrals), treated for instance in [FEF1]. We shall not discuss it here.]

Now let K be a Calderón-Zygmund kernel. The associated integral operator

$$T_K(\phi)(x) \equiv \int K(t)\phi(x - t)\, dt$$

makes no formal sense, even when ϕ is a testing function. This is so because K is not absolutely integrable. Instead we use the notion of the *Cauchy principal value* to evaluate this integral. To wit, let $\phi \in C_c^1(\mathbb{R}^N)$. Set

$$T_K(\phi)(x) = \text{P.V.} \int K(t)\phi(x - t)\, dt \equiv \lim_{\epsilon \to 0^+} \int_{|t| > \epsilon} K(t)\phi(x - t)\, dt.$$

Let us check that this limit actually exists. We write the integral in the right-hand side of the last formula as

$$\int_{1 > |t| > \epsilon} K(t)\phi(x - t)\, dt + \int_{|t| \geq 1} K(t)\phi(x - t)\, dt \equiv I + II.$$

Integral II converges trivially because the singularity of K at the origin is gone and the function ϕ is bounded and has compact support. For I, we write $K(x) = (|x|^N \cdot K(x))/|x|^N \equiv \Omega(x)/|x|^N$ (this is the classical Calderón-Zygmund notation). Notice that, by the $-N$-degree homogeneity of K,

$$\int_{1 > |t| > \epsilon} K(t)\, dt = \int_\epsilon^1 \int_{\Sigma_{N-1}} K(r\xi)\, d\sigma(\xi) r^{N-1} dr$$

$$= \int_\epsilon^1 \int_{\Sigma_{N-1}} \Omega(\xi) \cdot d\sigma(\xi) r^{-1} dr.$$

Now, by (5.2.1.3), the inner integral is 0. We see therefore that

$$\int_{1 > |t| > \epsilon} K(t)\, dt = 0.$$

As a result,

$$I = I(\epsilon) = \int_{1 > |t| > \epsilon} K(t)\phi(x - t)\, dt$$

$$= \int_{1 > |t| > \epsilon} K(t)[\phi(x - t) - \phi(x)]\, dt.$$

By the mean-value theorem, the expression in brackets is $\mathcal{O}(t)$ (see Appendix IX for this Landau notation). Therefore

$$|K(t)[\phi(x - t) - \phi(x)]| \le C \cdot r^{-N} \cdot r = Cr^{-N+1}.$$

[The term r^{-N} comes from the homogeneity of K, and the r is an estimate for $\mathcal{O}(t)$.] As a result, converting to polar coordinates, if $0 < \epsilon_1 < \epsilon_2$, then

$$|I(\epsilon_2) - I(\epsilon_1)| \le C \cdot \int_{\epsilon_1}^{\epsilon_2} r^{-N+1} \cdot r^{N-1} \, dr \le C \cdot [\epsilon_2 - \epsilon_1].$$

[The term r^{N-1} comes from the Jacobian of the change to polar coordinates.] In conclusion, the integral $I(\epsilon)$ now converges as $\epsilon \to 0^+$. Thus $T_K(\phi)$ makes sense pointwise (and in fact it defines T_K as a distribution, though we shall not explore that point here).

Building on ideas that we developed in the context of the Hilbert transform, the most fundamental question that we might now ask is whether

$$\|T_K(\phi)\|_{L^p} \le C \|\phi\|_{L^p}$$

for all $\phi \in C_c^1(\mathbb{R}^N)$. If, for some fixed p, this inequality holds, then a simple density argument would extend the operator T_K and the inequality to all $\phi \in L^p(\mathbb{R}^N)$. In fact, this inequality holds for $1 < p < \infty$ and fails for $p = 1, \infty$. The first of these two statements is called the *Calderón-Zygmund theorem*, and we prove it in Chapter 6. The second follows just as it did in Chapter 2 for the Hilbert transform; we leave the details as an exercise for the interested reader.

Here is a summary of what our discussion has revealed thus far:

Theorem 5.2.3 (Calderón-Zygmund) *Let K be a Calderón-Zygmund kernel. Then the operator*

$$T_K(\phi)(x) \equiv \text{P.V.} \int K(t)\phi(x - t) \, dt,$$

for $\phi \in C_c^1(\mathbb{R}^N)$, is well-defined. It is bounded in the L^p norm for $1 < p < \infty$. It is not bounded on L^1, nor is it bounded on L^∞.

It is natural to wonder whether there are spaces that are related to L^1 and L^∞, and which might serve as their substitutes for the purposes of singular integral theory. As we shall see, the correct approach is to consider that subspace of L^1 that behaves naturally under certain canonical singular integral operators. This approach yields a subspace of L^1 that is known as H^1 (or, more precisely, H_{Re}^1). The dual of this new space is a superspace of L^∞. It is called BMO (the functions of bounded mean oscillation). We shall explore these two new spaces, and their connections with the questions under discussion, as the chapter develops.

Notice that the discussion at the end of Section 2.2 on how to construct functions of a given homogeneity also tells us how to construct Calderón-Zygmund kernels. Namely, let ϕ be any smooth function on the unit sphere of $\mathbb{R}^N$ that integrates to zero with respect to area measure. Extend it to a function Ω on all of space (except the origin) so that Ω is homogeneous of degree zero, i.e., let $\Omega(x) = \phi(x/|x|)$. Then

$$K(x) = \frac{\Omega(x)}{|x|^N}$$

is a Calderón-Zygmund kernel.

5.3   An Aside on Integral Operators

One of the themes of twentieth-century harmonic analysis has been an effort to develop calculi of integral operators. The original goal was to develop a calculus of operators that would contain all parametrices (approximate right inverses) for elliptic partial differential operators. The idea of a "calculus" is to have operators T_k—with kernel k—so that when you need the adjoint of T_k you can say instantly what its kernel is (in terms of k). Or if you want to compose T_{k_1} with T_{k_2}, then you can say instantly what the kernel of the composition is (in terms of k_1, k_2). And likewise for the inverse of T_k, when that inverse exists.

One of the first attempts at generating such a calculus was to use the singular integral operators and the fractional integral operators as a generating set. This was clumsy, because compositions of these operators did not yield new operators of like type. Modern theories of pseudodifferential and Fourier integral operators have been much more successful, in part because we now realize that the properties of the adjoint, composition, and inverse that we sketched in the last paragraph are best achieved if we allow certain types of error terms consisting of "smoothing operators." Certainly any standard modern theory of pseudodifferential operators contains singular integrals and fractional integrals, but it also contains operators that are not translation-invariant, and it contains operators that can serve as our error terms.

In fact, one of the insights of Kohn and Nirenberg, who wrote the first paper [KON] developing a useful calculus of operators, is that it is more efficient to concentrate on the Fourier multiplier than on the kernel. Second, the calculus operates more smoothly if one classifies the operators in the calculus according to the decay of their symbols at infinity. An asymptotic expansion for the kernel comes later, and plays a less central role in the theory. A complete description of a calculus of pseudodifferential operators would take us far afield. The book [KRA3] contains an accessible introduction to the subject, as well as a list of many of the standard references.

5.4 A Look at Hardy Spaces in the Complex Plane

This section may be considered to be an optional look at the classical Hardy spaces—as developed by G. H. Hardy more than 70 years ago, based on earlier studies by O. Toeplitz—in complex function theory. The reader who chooses to skip this section will lose some motivation and context, but should be able to read the remainder of the book with little trouble.

Throughout this section we let $D \subseteq \mathbb{C}$ denote the open unit disc. For $0 < p < \infty$ we define

$$H^p(D) = \left\{ f \text{ holomorphic on } D: \sup_{0<r<1} \left[\frac{1}{2\pi} \int_0^{2\pi} |f(re^{i\theta})|^p d\theta \right]^{1/p} \right.$$

$$\left. \equiv \|f\|_{H^p} < \infty \right\}.$$

Also define

$$H^\infty(D) = \left\{ f \text{ holomorphic on } D : \sup_D |f| \equiv \|f\|_{H^\infty} < \infty \right\}.$$

We record in passing a useful result that is based on the fact that $|f|^p$ is subharmonic when f is holomorphic and $0 < p < \infty$. We refer the reader to [KRA4, Section 8.1] for a proof.

Lemma 5.4.1 *Let $f \in H^p(D)$, $0 < p < \infty$. Let $0 < r_1 < r_2 < 1$. Then*

$$\int_0^{2\pi} |f(r_1 e^{i\theta})|^p \, d\theta \leq \int_0^{2\pi} |f(r_2 e^{i\theta})|^p \, d\theta.$$

The fundamental result in the subject of H^p, or *Hardy* spaces (and also the fundamental result of this section), is that if $f \in H^p(D)$, then the limit

$$\lim_{r \to 1^-} f(re^{i\theta}) \equiv f^*(e^{i\theta})$$

exists for almost every $\theta \in [0, 2\pi)$. For $1 \leq p \leq \infty$, the function f can be recovered from f^* by way of the Cauchy or Poisson integral formulas; for $p < 1$ this "recovery" process is more subtle and must proceed by way of distributions. Once this pointwise boundary limit result is established, then an enormous and rich mathematical structure unfolds (see [KAT], [HOF], [GAR]).

Recall (see [GRK], [KRA4], and Sections 1.3 and 4.1) that the Poisson kernel for the disc is

$$P_r(e^{i\theta}) = \frac{1 - r^2}{1 - 2r\cos\theta + r^2}.$$

Let, for $1 \le p < \infty$,

$$\mathbf{h}^p(D) = \left\{ f \text{ harmonic on } D : \sup_{0<r<1} \left[\frac{1}{2\pi} \int_0^{2\pi} |f(re^{i\theta})|^p d\theta \right]^{1/p} \right.$$

$$\left. \equiv \|f\|_{\mathbf{h}^p} < \infty \right\}$$

Also let

$$\mathbf{h}^\infty(D) = \left\{ f \text{ harmonic on } D : \sup_D |f| \equiv \|f\|_{\mathbf{h}^\infty} < \infty \right\}.$$

Proposition 5.4.2 *Let* $1 < p \le \infty$ *and* $f \in \mathbf{h}^p(D)$. *Then there is an* $f^* \in L^p(\partial D)$ *such that, for all* $re^{i\theta} \in D$,

$$f(re^{i\theta}) = \int_0^{2\pi} f^*(e^{i\psi}) P_r(e^{i(\theta-\psi)}) d\psi.$$

Proof. Define $f_r(e^{i\theta}) = f(re^{i\theta}), 0 < r < 1$. Then $\{f_r\}_{0<r<1}$ is a bounded subset of $L^p(\partial D) = (L^{p'}(\partial D))^*$, $p' = p/(p-1)$ (Theorem 0.3.9). By the Banach-Alaoglu theorem (Appendix VI or [RUD3]), there is a subsequence f_{r_j} that converges weak-* to some f^* in $L^p(\partial D)$. For any fixed $0 < r < 1$, let $r < r_j < 1$. Then

$$f(re^{i\theta}) = f_{r_j}\left((r/r_j)e^{i\theta}\right) = \int_0^{2\pi} f_{r_j}(e^{i\psi}) P_{r/r_j}\left(e^{i(\theta-\psi)}\right) d\psi$$

because $f_{r_j} \in C(\overline{D})$ and f_{r_j} is harmonic in D. Now $P_{r/r_j} \in C(\partial D) \subseteq L^{p'}(\partial D)$. We render the right-hand side of the last equation as

$$\int_0^{2\pi} f_{r_j}(e^{i\psi}) P_r(e^{i(\theta-\psi)}) d\psi$$

$$+ \int_0^{2\pi} f_{r_j}(e^{i\psi}) \left[P_{r/r_j}(e^{i(\theta-\psi)}) - P_r(e^{i(\theta-\psi)}) \right] d\psi.$$

For $0 < r < 1$ fixed, $j \to \infty$, the second expression vanishes (by Hölder, and because the expression in brackets converges uniformly to 0), and the first integral tends to

$$\int_0^{2\pi} f^*(e^{i\psi}) P_r(e^{i(\theta-\psi)}) d\psi$$

by weak-$*$ convergence. This is the desired result. $\qquad\square$

Remark. The proof breaks down for $p = 1$ since L^1 is not the dual of any Banach space. This breakdown is not merely a failure of technique: the harmonic function

$$f(re^{i\psi}) = P_r(e^{i\psi})$$

satisfies

$$\sup_{0<r<1} \int_0^{2\pi} |f(re^{i\theta})| d\theta < \infty,$$

but f is not the Poisson integral of any L^1 function, as follows from the next proposition and the fact that $\lim_{r\to 1^-} P_r(e^{i\psi}) = 0$ for all $\psi \in (0, 2\pi)$. See also the remark following the proposition.

Proposition 5.4.3 *Let* $f \in L^p(\partial D), 1 \le p < \infty$. *Then* $\lim_{r\to 1^-} P_r f = f$ *in the* L^p *norm.*

Remark. The result is false for $p = \infty$ if f is discontinuous. The correct analogue in the uniform case is that if $f \in C(\partial D)$, then $P_r f \to f$ uniformly.

As an exercise, consider a Borel measure μ on ∂D. Show that its Poisson integral $P_r \mu$ converges in the weak-$*$ topology to μ. $\qquad\square$

Proof of Proposition 5.4.3. If $f \in C(\partial D)$, then the result is clear from the integral form of the solution of the Dirichlet problem. If $f \in L^p(\partial D)$ is arbitrary, let $\epsilon > 0$ and choose $g \in C(\partial D)$ such that $\|f - g\|_{L^p} < \epsilon$. Then

$$\|P_r f - f\|_{L^p} \leq \|P_r(f - g)\|_{L^p} + \|P_r g - g\|_{L^p} + \|g - f\|_{L^p}$$
$$\leq \|P_r\|_{L^1}\|f - g\|_{L^p} + \|P_r g - g\|_{L^p} + \epsilon$$
$$\leq \epsilon + o(1) + \epsilon$$

as $r \to 1^-$. [Note the use of Landau's notation—see Appendix IX.]

$\square$

Remark. For an alternative proof of 5.4.3, observe that FAPI shows easily that the limit exists in 5.4.3. An extra argument is needed to identify the limit as f itself. $\square$

We now use the Hardy-Littlewood maximal function, particularly its control of the Poisson kernel (see Section 1.5), to prove a result about boundary limits of h^p functions. The astute reader will notice that the question treated in Section 1.5 (about summation of Fourier series) and the question treated here about boundary behavior are really the same question. The proof is in the proof. [See also the discussion of the Hilbert transform from the point of view of analytic function theory at the end of Section 1.6.]

Theorem 5.4.4 *Let* $f \in \mathbf{h}^p(D)$ *and* $1 < p \leq \infty$. *Let* f^* *be as in Proposition 5.4.2. Then*

$$\lim_{r \to 1^-} f(re^{i\theta}) = f^*(e^{i\theta}), \quad a.e. \ e^{i\theta} \in \partial D.$$

Proof. Because $\mathbf{h}^\infty \subseteq \mathbf{h}^p$ for all p, it suffices to handle the case $p < \infty$ and f real-valued.

We note that if g is continuous on $\mathbb{T}$, then the Poisson integral $P_r g$ is continuous on the closure of the disc, and the assertion we are making is trivially true at every point. Also, the space of continuous functions is dense in $L^p(\mathbb{T})$. Finally, the maximal function manufactured from the P_r—

$$P^* f(e^{it}) = \sup_{0 < r < 1} |P_r f(e^{it})|$$

—is weak-type $(1, 1)$ because[1] it is majorized by the Hardy-Littlewood maximal function (see Theorem 1.5.4). Also P^* maps L^∞ to L^∞. By Marcinkiewicz interpolation (Appendix III), P^* is bounded on L^p for $1 < p < \infty$.

We may now apply FAPII to draw the conclusion of the theorem.

$\square$

The informal statement of Theorem 5.4.4 is that f has radial boundary limits almost everywhere.

Theorem 5.4.4 contains essentially all that can be said about the range of p for which p^{th}-power-integrable harmonic functions have boundary values. Why are holomorphic functions better behaved? The classical method for answering this question is to use Blaschke factorization:

Definition 5.4.5 For $a \in \mathbb{C}$, $|a| < 1$, the *Blaschke factor* at a is

$$B_a(z) = \frac{z - a}{1 - \overline{a}z}.$$

It is elementary to verify that B_a is holomorphic on a neighborhood of $\overline{D}$ and that $|B_a(e^{i\theta})| = 1$ for all $0 \leq \theta < 2\pi$. Indeed, if $|z| = 1$ then

$$|B_a(z)| = \left| \frac{\overline{z}}{1} \cdot \frac{z - a}{1 - \overline{a}z} \right| = \left| \frac{1 - \overline{z}a}{1 - \overline{a}z} \right| = 1.$$

---

[1] That $P^* f$ is measurable is seen by the trick that we used in Section 1.5 to verify the measurability of the Hardy-Littlewood operator: It suffices to take the supremum on the right over only rational r. And the supremum of countably many measurable functions is measurable.

Lemma 5.4.6 *Let $0 < r < 1$ and suppose that f is holomorphic on a neighborhood of $\overline{D}(0, r)$. Let $p_1, \ldots, p_k$ be the zeros of f (listed with multiplicity) in $D(0, r)$. Assume that $f(0) \neq 0$ and that $f(re^{it}) \neq 0$, all t. Then*

$$\log |f(0)| + \log \prod_{j=1}^{k} r|p_j|^{-1} = \frac{1}{2\pi} \int_0^{2\pi} \log |f(re^{it})| \, dt.$$

Proof. The function

$$F(z) = \frac{f(z)}{\prod_{j=1}^{k} B_{p_j/r}(z/r)}$$

is holomorphic and zero-free on a neighborhood of $\overline{D}(0, r)$; hence $\log |F|$ is harmonic on a neighborhood of $\overline{D}(0, r)$. Thus

$$\log |F(0)| = \frac{1}{2\pi} \int_0^{2\pi} \log |F(re^{it})| \, dt$$

$$= \frac{1}{2\pi} \int_0^{2\pi} \log |f(re^{it})| \, dt,$$

which means that

$$\log |f(0)| + \log \prod_{j=1}^{k} \frac{r}{|p_j|} = \frac{1}{2\pi} \int_0^{2\pi} \log |f(re^{it})| \, dt. \qquad \square$$

Notice that, by the continuity of the integral as a function of r, Lemma 5.4.6 holds even if f has zeros on $\{re^{it}\}$.

Corollary 5.4.7 *If f is holomorphic in a neighborhood of $\overline{D}(0, r)$, then*

$$\log |f(0)| \leq \frac{1}{2\pi} \int_0^{2\pi} \log |f(re^{it})| \, dt.$$

Proof. The term $\log \prod_{j=1}^{k} |r/p_j|$ in Lemma 5.4.6 is positive. $\qquad \square$

Corollary 5.4.8 *For $\alpha > 0$, let $\log^+ \alpha \equiv \max\{\log \alpha, 0\}$. If f is holomorphic on D, $f(0) \neq 0$, and $\{p_1, p_2, \ldots\}$ are the zeros of f in D counting multiplicities, then*

$$\log|f(0)| + \log \prod_{j=1}^{\infty} \frac{1}{|p_j|} \leq \sup_{0<r<1} \frac{1}{2\pi} \int_0^{2\pi} \log^+ |f(re^{it})| dt.$$

Proof. Apply Lemma 5.4.6, letting $r \to 1^-$. $\qquad\qquad\qquad\qquad\square$

Corollary 5.4.9 *If $f \in H^p(D)$, $0 < p \leq \infty$, and $\{p_1, p_2, \ldots\}$ are the zeros of f counting multiplicities, then $\sum_{j=1}^{\infty}(1 - |p_j|) < \infty$.*

Proof. Since f vanishes to finite order k at 0, we may replace f by $f(z)/z^k$ and assume that $f(0) \neq 0$. Since $H^\infty \subseteq H^p$ for all p, we may take p to be finite. Notice that, if $0 < p < \infty$, then

$$\int \log^+ |f| \leq C_p \int (1 + |f|)^p.$$

The constant C_p depends, of course, on p.

It follows from Corollary 5.4.8 and the hypothesis that $f \in H^p(D)$ that

$$\log \prod_{j=1}^{\infty} \frac{1}{|p_j|} < \infty$$

so that $\prod(1/|p_j|)$ converges; hence $\prod |p_j|$ converges. Therefore $\sum_j (1 - |p_j|) < \infty$ (see [GRK] for basic facts about infinite products). $\qquad\qquad\qquad\qquad\square$

Proposition 5.4.10 *If $\{p_1, p_2, \ldots\} \subseteq D$ satisfy $\sum_j (1 - |p_j|) < \infty$, $p_j \neq 0$ for all j, then*

$$\prod_{j=1}^{\infty} \frac{-\overline{p_j}}{|p_j|} B_{p_j}(z)$$

converges uniformly on compact subsets of D.

Proof. Restrict attention to $|z| \leq r < 1$. Then the assertion that the infinite product converges uniformly on the disc $\overline{D}(0, r)$ is equivalent (see [GRK]) to the assertion that

$$\sum_j \left| 1 + \frac{\overline{p}_j}{|p_j|} B_{p_j}(z) \right|$$

converges uniformly there. But, for such z,

$$
\begin{aligned}
\left| 1 + \frac{\overline{p}_j}{|p_j|} B_{p_j}(z) \right| &= \left| \frac{|p_j| - |p_j|\overline{p}_j z + \overline{p}_j z - |p_j|^2}{|p_j|(1 - z\overline{p}_j)} \right| \\
&= \left| \frac{(|p_j| + z\overline{p}_j)(1 - |p_j|)}{|p_j|(1 - z\overline{p}_j)} \right| \\
&\leq \frac{(1 + r)(1 - |p_j|)}{1 - r},
\end{aligned}
$$

so the convergence is uniform. $\square$

Definition 5.4.11 Let $0 < p \leq \infty$ and $f \in H^p(D)$. Assume that f is not identically zero. By dividing out a power of z, we may assume that f does not vanish at the origin. Let $\{p_1, p_2, \ldots\}$ be the zeros of f counted according to multiplicities. Let

$$B(z) = \prod_{j=1}^{\infty} \frac{-\overline{p}_j}{|p_j|} B_{p_j}(z).$$

Then B is a well-defined holomorphic function on D by Proposition 5.4.10. Let $F = f/B$. By the Riemann removable singularities theorem, F is a well-defined, nonvanishing holomorphic function on D. The representation $f = F \cdot B$ is called the *canonical factorization* of f.

Proposition 5.4.12 *Let $f \in H^p(D), 0 < p \leq \infty$, and assume that $f(0) \neq 0$. Let $f = F \cdot B$ be its canonical factorization. Then $F \in H^p(D)$ and $\|F\|_{H^p} = \|f\|_{H^p}$.*

Proof. Trivially, $|F| = |f/B| \geq |f|$, so $\|F\|_{H^p} \geq \|f\|_{H^p}$. For $K = 1, 2, \ldots,$ let

$$B_K(z) = \prod_{j=1}^{K} \frac{-\overline{p}_j}{|p_j|} B_{p_j}(z).$$

Let $F_K = f/B_K$. Since $|B_K(e^{it})| = 1$, for all t, we have $\|F_K\|_{H^p} = \|f\|_{H^p}$ (use Lemma 5.4.1 and the fact that $B_K(re^{it}) \to B_K(e^{it})$ uniformly in t as $r \to 1^-$). If $0 < r < 1$, then

$$\left[\frac{1}{2\pi} \int_0^{2\pi} |F(re^{it})|^p dt \right]^{1/p} = \lim_{K \to \infty} \left[\int_0^{2\pi} |F_K(re^{it})|^p dt \right]^{1/p}$$

$$\leq \lim_{K \to \infty} \|F_K\|_{H^p} = \|f\|_{H^p}.$$

Therefore $\|F\|_{H^p} \leq \|f\|_{H^p}$.   □

Corollary 5.4.13 *If* $\{p_1, p_2, \ldots\}$ *is a sequence of points in* D *satisfying* $\sum_j (1 - |p_j|) < \infty$ *and if* $B(z) = \prod_j (-\overline{p}_j/|p_j|) B_{p_j}(z)$ *is the corresponding Blaschke product, then*

$$B^*(e^{it}) \equiv \lim_{r \to 1^-} B(re^{it})$$

exists and has modulus 1 almost everywhere.

Proof. The conclusion that the limit exists follows from Theorem 5.4.4 and the fact that $B \in H^\infty \subseteq \mathbf{h}^\infty$. For the other assertion, note that the canonical factorization for B is $B = 1 \cdot B$. Therefore, by Proposition 5.4.12,

$$\left[\frac{1}{2\pi} \int |B^*(e^{it})|^2 dt \right]^{1/2} = \|B\|_{H^2} = \|1\|_{H^2} = 1;$$

hence $|B^*(e^{it})| = 1$ almost everywhere.   □

Theorem 5.4.14 *If* $f \in H^p(D)$ *and* $0 < p \leq \infty$, *then*

$$\lim_{r \to 1^-} f(re^{i\theta})$$

exists for almost every $e^{i\theta} \in \partial D$ and equals $f^*(e^{i\theta})$. Also, $f^* \in L^p(\partial D)$ and

$$\|f^*\|_{L^p} = \|f\|_{H^p} \equiv \sup_{0<r<1} \left[\frac{1}{2\pi} \int_0^{2\pi} |f(re^{i\theta})|^p d\theta \right]^{1/p}.$$

Proof. Use Definition 5.4.11 to write $f = B \cdot F$, where $F \in H^p$ has no zeros and B is a Blaschke product. Then $F^{p/2}$ is a well-defined function in $H^2 \subseteq \mathbf{h}^2$ and thus has (by 5.4.4) the appropriate boundary values almost everywhere. *A fortiori* (but beware of ambiguities caused by multiples of 2π!), F has radial boundary limits almost everywhere. Since $B \in H^\infty$, B has radial boundary limits almost everywhere. It follows that f does as well. The final assertion follows from the corresponding fact for H^2 functions (exercise). □

5.5   The Real-Variable Theory of Hardy Spaces

We saw in Section 5.4 that a function f in the Hardy class $H^p(D)$ on the disc may be identified in a natural way with its boundary function, which we continue to call f^*. Fix attention for the moment on $p = 1$.

If $\phi \in L^1(\partial D)$ and is real-valued, then we may define a harmonic function u on the disc by

$$u(re^{i\theta}) = \frac{1}{2\pi} \int_0^{2\pi} \phi(e^{i\psi}) \frac{1 - r^2}{1 - 2r\cos(\theta - \psi) + r^2} d\psi.$$

Of course this is just the usual Poisson integral of ϕ. As was proved in Section 5.4, the function ϕ is the "boundary function" of u in a natural manner. Let v be the harmonic conjugate of u on the disc (we may make the choice of v unique by demanding that $v(0) = 0$). Thus $h \equiv u + iv$ is holomorphic. We may then ask whether the function v has a boundary limit function $\widetilde{\phi}$.

To see that $\widetilde{\phi}$ exists, we reason as follows: Suppose that the original function ϕ is nonnegative (any real-valued ϕ is the difference of two such functions, so there is no loss in making this additional hypothesis). Since the Poisson kernel is positive, it follows that $u > 0$. Now consider the holomorphic function

$$F = e^{-u-iv}.$$

The positivity of u implies that F is bounded. Thus $F \in H^\infty$. By Theorem 5.4.14, we may conclude that F has radial boundary limits at almost every point of ∂D. Unraveling our notation (and thinking a moment about the ambiguity caused by multiples of 2π), we find that v itself has radial boundary limits almost everywhere. We define thereby the function $\widetilde{\phi}$.

Of course the function h can be expressed (up to an additive factor of $1/2$ and a multiplicative factor of $1/2$—see our calculations at the end of Section 4.1) as the Cauchy integral of ϕ. The real part of the Cauchy kernel is the Poisson kernel (again up to a multiplicative and additive factor of $1/2$—see the calculation following, so it makes sense that the real part of h on D converges back to ϕ. By the same token, the imaginary part of h is the integral of ϕ against the imaginary part of the Cauchy kernel, and it will converge to $\widetilde{\phi}$. It behooves us to calculate the imaginary part of the Cauchy kernel.

In the Cauchy integrand $(1/i)[1/(\zeta - z)]d\zeta$ we set $z = re^{i\theta}$ and $\zeta = e^{i\psi}$. Then

$$
\begin{aligned}
\frac{1}{i}\frac{1}{\zeta - z}\,d\zeta &= \frac{-i\overline{\zeta}\,d\zeta}{\overline{\zeta}(\zeta - z)} \\
&= \frac{d\psi}{1 - re^{i(\theta-\psi)}} \\
&= \frac{(1 - re^{-i(\theta-\psi)})d\psi}{|1 - re^{i(\theta-\psi)}|^2} \\
&= \frac{[1 - r\cos(\theta - \psi)] + i[r\sin(\theta - \psi)]}{|1 - re^{i(\theta-\psi)}|^2}d\psi. \quad (5.5.1)
\end{aligned}
$$

For convenience, let us set $t = \theta - \psi$. Then the imaginary part of the Cauchy kernel, as just normalized and calculated, equals

$$\widetilde{P}_r(e^{it}) \equiv \frac{r \sin t}{1 - 2r \cos t + r^2}.$$

If we denote the real part of (5.5.1) by P_r^*, then a quick calculation shows that $P_r^* - 1/2 = (1/2) \cdot P_r$, where P_r is the usual Poisson kernel. We conclude that

$$v(re^{i\theta}) = 2 \cdot \widetilde{P}_r * \phi(e^{i\theta}).$$

If we formally let $r \to 1^-$ (and suppress some nasty details—see [KAT]), then we find that $\widetilde{\phi}$ is just the convolution of ϕ with the kernel

$$\begin{aligned}
\widetilde{k}(t) &\equiv 2 \cdot \frac{\sin t}{2 - 2 \cos t} \\
&= 2 \cdot \frac{2 \sin t/2 \cos t/2}{2[1 - \cos^2 t/2 + \sin^2 t/2]} \\
&= \cot(t/2).
\end{aligned}$$

The operator

$$H : \phi \longmapsto \phi * \cot(t/2)$$

is the Hilbert transform that we studied in Chapter 1, where it was motivated by questions of norm convergence of classical Fourier series. Part of the reason for the importance of the Hilbert transform is that it is the unique translation-invariant singular integral operator in dimension 1. Another part of the reason is that it arises so naturally in many different contexts—Fourier series and complex analysis are just two of them.

And our calculations will now give us a new way to think about the Hardy space $H^1(D)$. For if ϕ and $\widetilde{\phi}$ are, respectively, the boundary functions of Re f and Im f for an $f \in H^1$, then $\phi, \widetilde{\phi} \in L^1$, and our preceding discussion shows that (up to our usual correction factors)

$\widetilde{\phi} = H\phi$. *But this relationship is worth special note:* We have already proved that the Hilbert transform is not bounded on L^1, yet we see that the functions ϕ that arise as boundary functions of the real parts of functions in H^1 have the property that $\phi \in L^1$ and (surprisingly) $H\phi \in L^1$. These considerations motivate the following *real-variable definition of the Hardy space* H^1:

Definition 5.5.2 A function $f \in L^1$ (on the circle, or on the real line) is said to be in the *real-variable Hardy space* H^1_{Re} if the Hilbert transform of f is also in L^1. The H^1_{Re} norm of f is given by

$$\|\phi\|_{H^1_{\mathrm{Re}}} \equiv \|\phi\|_{L^1} + \|H\phi\|_{L^1}.$$

In higher dimensions the role of the Hilbert transform is played by a family of singular integral operators. On $\mathbb{R}^N$, let

$$K_j(x) = \frac{x_j/|x|}{|x|^N} \qquad , \; j = 1, \ldots, N.$$

Notice that each K_j possesses the three defining properties of a Calderón-Zygmund kernel: it is smooth away from the origin, homogeneous of degree $-N$, and satisfies the mean-value condition because $\Omega_j(x) \equiv x_j/|x|$ is odd. We set $R_j f$ equal to the singular integral operator with kernel K_j applied to f, and call this operator the jth *Riesz transform*. Since the kernel of the jth Riesz transform is homogeneous of degree $-N$, it follows that the Fourier multiplier for this operator is homogeneous of degree zero (see Proposition 2.2.9). It is possible to calculate (though we shall not do it) that this multiplier has the form $c \cdot \xi_j/|\xi|$ (see [STE1] for the details). Observe that on $\mathbb{R}^1$ there is just one Riesz transform, and it is the Hilbert transform.

It turns out that the N-tuple $(K_1(x), \ldots K_N(x))$ behaves naturally with respect to rotations, translations, and dilations in just the same way that the Hilbert kernel $1/t$ does in $\mathbb{R}^1$ (see [STE1] for details). These considerations, and the ideas leading up to the preceding definition, give us the following:

Definition 5.5.3 (Stein-Weiss [STG2]) Let $f \in L^1(\mathbb{R}^N)$. We say that f is in the real-variable Hardy space of order 1, and write $f \in H^1_{\mathrm{Re}}$, if $R_j f \in L^1$, $j = 1, \ldots, N$. The norm on this new space is

$$\|f\|_{H^1_{\mathrm{Re}}} \equiv \|f\|_{L^1} + \|R_1 f\|_{L^1} + \cdots + \|R_N f\|_{L^1}.$$

We have provided a motivation for this last definition by way of the theory of singular integrals. It is also possible to provide a motivation via the Cauchy-Riemann equations. We now explain:

Recall that, in the classical complex plane, the Cauchy-Riemann equations for a C^1, complex-valued function $f = u + iv$ are

$$\frac{\partial u}{\partial x} = \frac{\partial v}{\partial y}$$

$$\frac{\partial u}{\partial y} = -\frac{\partial v}{\partial x}.$$

The function f will satisfy this system of two linear first-order equations if and only if f is holomorphic (see [GRK] for details).

On the other hand, if (v, u) is the gradient of a harmonic function h then

$$\frac{\partial u}{\partial x} = \frac{\partial^2 h}{\partial x \partial y} = \frac{\partial^2 h}{\partial y \partial x} = \frac{\partial v}{\partial y}$$

and

$$\frac{\partial u}{\partial y} = \frac{\partial^2 h}{\partial y^2} = -\frac{\partial^2 h}{\partial x^2} = -\frac{\partial v}{\partial x}.$$

[Note how we have used the fact that $\triangle h = 0$.] These are the Cauchy-Riemann equations. One may also use elementary ideas from multi-variable calculus (see [GRK]) to see that a pair (v, u) that satisifies the Cauchy-Riemann equations in this way must be the gradient of a harmonic function.

Passing to N variables, let us now consider a function $f \in H^1_{\mathrm{Re}}(\mathbb{R}^N)$. Set $f_0 = f$ and $f_j = R_j f$, $j = 1, \ldots, N$. By definition,

$f_j \in L^1(\mathbb{R}^N)$, $j = 1, \ldots, N$. Thus it makes sense to consider

$$u_j(x, y) = P_y * f_j(x), \quad j = 0, 1, \ldots, N,$$

where

$$P_y(x) \equiv c_N \frac{y}{(|x|^2 + y^2)^{(N+1)/2}}$$

is the standard Poisson kernel for the upper half-space

$$\mathbb{R}_+^N \equiv \{(x, y) : x \in \mathbb{R}^N, y > 0\}.$$

A formal calculation (see [STG1]) shows that

$$\frac{\partial u_j}{\partial x_k} = \frac{\partial u_k}{\partial x_j}$$

for $j, k = 0, 1, \ldots, N$ and

$$\sum_{j=1}^{N} \frac{\partial u_j}{\partial x_j} = 0.$$

These are the *generalized Cauchy-Riemann equations*. The two conditions taken together are equivalent to the hypothesis that the $(N + 1)$-tuple $(u_0, u_1, \ldots, u_N)$ be the gradient of a harmonic function h on $\mathbb{R}_+^{N+1}$. See [STG1] for details.

Both the singular integrals point of view and the Cauchy-Riemann equations point of view can be used to define H_{Re}^p for $0 < p < 1$. These definitions, however, involve algebraic complications that are best avoided in the present book. [Details may be found in [FES] and [STG2].] In the next section we present another point of view for Hardy spaces that treats all values of p, $0 < p \leq 1$, simultaneously.

We close this section with an application of the Riesz transforms to an imbedding theorem for function spaces. This is a version of the so-called *Sobolev imbedding theorem*.

Theorem 5.5.4 *Fix a dimension $N > 1$ and let $1 < p < N$. Let $f \in L^p(\mathbb{R}^N)$ with the property that $(\partial/\partial x_j)f$ exists and is in L^p, $j = 1, \ldots, N$. Then $f \in L^q(\mathbb{R}^N)$, where $1/q = 1/p - 1/N$.*

Proof. As usual, we content ourselves with a proof of an *a priori* inequality for $f \in C_c^\infty(\mathbb{R}^N)$. We write

$$\widehat{f}(\xi) = \sum_{j=1}^{N} \frac{1}{|\xi|} \cdot \frac{\xi_j}{|\xi|} \cdot [\xi_j \widehat{f}(\xi)].$$

Observe that $\xi_j \widehat{f}(\xi)$ is (essentially) the Fourier transform of $(\partial/\partial x_j)f$. Also $\xi_j/|\xi|$ is the multiplier for the jth Riesz transform. And $1/|\xi|$ is the Fourier multiplier for a fractional integral. [All of these statements are true up to constant multiples, which we omit.] Using operator notation, we may therefore rewrite the last displayed equation as

$$f(x) = \sum_{j=1}^{N} \mathcal{I}^1 \circ R_j \left(\frac{\partial}{\partial x_j} f(x) \right).$$

We know by hypothesis that $(\partial/\partial x_j)f \in L^p$. Now R_j maps L^p to L^p and I_1 maps L^p to L^q, where $1/q = 1/p - 1/N$ (see Theorem 5.1.3). That completes the proof. $\qquad\square$

5.6 The Maximal-Function Characterization of Hardy Spaces

Recall (see Section 1.5) that the classical Hardy-Littlewood maximal function

$$Mf(x) \equiv \sup_{r>0} \frac{1}{m[B(x,r)]} \int_{B(x,r)} |f(t)| \, dt$$

is not bounded on L^1. Part of the reason for this failure is that L^1 is not a propitious space for harmonic analysis, and another part of the reason is that the characteristic function of a ball is not smooth. To understand this last remark, we set $\phi = [1/m(B(0,1)] \cdot \chi_{B(0,1)}$, the normalized characteristic function of the unit ball, and note that

$$Mf(x) = \sup_{R>0} \left|(\alpha^R\phi) * f(x)\right|$$

(where α^R is the dilation operator defined in Section 2.2).[2] It is natural to ask what would happen if we were to replace $\phi = [1/m(B(0,1)] \cdot \chi_{B(0,1)}$ in the definition of Mf with a smooth testing function ϕ. This we now do.

We fix a function $\phi_0 \in C_c^\infty(\mathbb{R}^N)$ and, for technical reasons, we assume that $\int \phi_0\, dx = 1$. We define

$$f^*(x) = \sup_{R>0} \left|(\alpha^R\phi_0) * f(x)\right|$$

for $f \in L^1(\mathbb{R}^N)$. We say that $f \in H^1_{\max}(\mathbb{R}^N)$ if $f^* \in L^1$. The following theorem, whose complicated proof we omit (but see [FES] or [STE2]), justifies this new definition:

Theorem 5.6.1 Let $f \in L^1(\mathbb{R}^N)$. Then $f \in H^1_{\mathrm{Re}}(\mathbb{R}^N)$ if and only if $f \in H^1_{\max}(\mathbb{R}^N)$.

It is often said that, in mathematics, a good theorem will spawn an important new definition. That is what will happen for us right now:

Definition 5.6.2 Let $f \in L^1_{\mathrm{loc}}(\mathbb{R}^N)$ and $0 < p \le 1$. We say that $f \in H^p_{\max}(\mathbb{R}^N)$ if $f^* \in L^p$.

It turns out that this definition of H^p is equivalent to the definitions using singular integrals or Cauchy-Riemann equations that we alluded to, but did not enunciate, at the end of the last section. For convenience, we take Definition 5.6.2 to be our definition of H^p when $p < 1$. In the next section we shall begin to explore what has come to

---

[2] Observe that there is no loss of generality, and no essential change, in omitting the absolute values around f that were originally present in the definition of M. For if M is restricted to positive f, then the usual maximal operator results.

be considered the most flexible approach to Hardy spaces. It has the advantage that it requires a minimum of machinery, and can be adapted to a variety of situations—boundaries of domains, manifolds, Lie groups, and other settings as well. This is the so-called *atomic theory* of Hardy spaces.

5.7 The Atomic Theory of Hardy Spaces

We first formulate the basic ideas concerning atoms for $p = 1$. Then we shall indicate the generalization to $p < 1$. The complete story of the atomic theory of Hardy spaces may be found in [STE2]. Foundational papers in the subject are [COI] and [LAT].

Let $a \in L^1(\mathbb{R}^N)$. We impose three conditions on the function a:

(5.7.1) The support of a lies in some ball $B(x, r)$;

(5.7.2) We have the estimate

$$|a(t)| \le \frac{1}{m(B(x, r))}$$

for every t.

(5.7.3) We have the mean-value condition

$$\int a(t)\, dt = 0.$$

A function a that enjoys these three properties is called a 1-*atom*.

Notice the mean-value-zero property in Axiom (5.7.3). Assuming that atoms are somehow basic or typical H^1 functions (and this point we shall treat momentarily), we might have anticipated this vanishing-moment condition as follows. Let $f \in H^1(\mathbb{R}^N)$ according to the classical definition using Riesz transforms. Then $f \in L^1$ and $R_j f \in L^1$ for each j. Taking Fourier transforms, we see that

$$\widehat{f} \in C_0 \quad \text{and} \quad \widehat{R_j f}(\xi) = c\frac{\xi_j}{|\xi|}\,\widehat{f}(\xi) \in C_0, \qquad j = 1, \dots, N.$$

The only way that the last N conditions could hold—in particular that $[\xi_j/|\xi|] \cdot \widehat{f}(\xi)$ could be continuous at the origin—is for $\widehat{f}(0)$ to be 0. But this says that $\int f(t)\, dt = 0$. That is the mean-value-zero condition that we are now mandating for an atom.

Now let us discuss p-atoms for $0 < p < 1$. It turns out that we must stratify this range of p's into infinitely many layers, and treat each layer separately. Fix a value of p, $0 < p \leq 1$. Let a be a measurable function. We impose three conditions for a to be a p-atom:

($\mathbf{5.7.4_p}$) The support of a lies in some ball $B(x, r)$;

($\mathbf{5.7.5_p}$) We have the estimate

$$|a(t)| \leq \frac{1}{m(B(x, r))^{1/p}} \qquad \text{for all } t.$$

($\mathbf{5.7.6_p}$) We have the mean-value condition

$$\int a(t) \cdot t^\beta\, dt = 0$$

for all multi-indices β with $|\beta| \leq N \cdot (p^{-1} - 1)$.

The aforementioned stratification of values of p now becomes clear: if k is a nonnegative integer then, when

$$\frac{N}{N + k + 1} < p \leq \frac{N}{N + k},$$

we demand that a p-atom a have vanishing moments up to and including order k. This means that the integral of a against any monomial of degree less than or equal to k must be zero.

The basic fact about the atomic theory is that a p-atom is a "typical" H_{Re}^p function. More formally:

Theorem 5.7.7 *Let* $0 < p \leq 1$. *For each* $f \in H_{\text{Re}}^p(\mathbb{R}^N)$ *there exist* p-atoms a_j *and complex numbers* β_j *such that*

$$f = \sum_{j=1}^{\infty} \beta_j a_j \qquad (5.7.7.1)$$

and the sequence of numbers $\{\beta_j\}$ satisfies $\sum_j |\beta_j|^p < \infty$. The sense in which the series representation (5.7.7.1) for f converges is a bit subtle (that is, it involves distribution theory) when $p < 1$, and we shall not discuss it here. When $p = 1$, the convergence is in L^1.

The converse to the decomposition (5.7.7.1) holds as well: Any sum as in (5.7.7.1) represents an $H^p_{\mathrm{Re}}(\mathbb{R}^N)$ function (where we may take this space to be defined by any of the preceding definitions).

If one wants to study the action of a singular integral operator, or a fractional integral operator, on H^p, then by linearity it suffices to check the action of that operator on an atom.

One drawback of the atomic theory is that a singular integral operator will not generally send atoms to atoms. Thus the program described in the last paragraph is not quite as simple as it sounds. To address this problem, a theory of "molecules" has been invented. Just as the name suggests, a p-molecule is an agglomeration of atoms—subject to certain rules. And it is a theorem that a singular integral operator will map molecules to molecules. See [TAW] for further details.

In Chapter 6, when we introduce spaces of homogeneous type, we shall continue our development of the atomic theory in a more general setting.

As an exercise, the reader may wish to consider what space of functions is obtained when the mean-value condition (Axiom (5.7.3)) is omitted from the definition of 1-atom. Of course the resulting space is L^1, and this will continue to hold in Chapter 6 when we are in the more general setting of spaces of homogeneous type. Matters are more complicated for either $p < 1$ or $p > 1$.

5.8 Ode to BMO

The space of functions of bounded mean oscillation was first treated by F. John and L. Nirenberg (see [JON]) in their study of certain nonlinear partial differential equations that arise in the study of minimal surfaces. Their ideas were, in turn, inspired by deep ideas of J. Moser [MOS1], [MOS2].

A function $f \in L^1_{\text{loc}}(\mathbb{R}^N)$ is said to be in BMO (the functions of *bounded mean oscillation*) if

$$\|f\|_* \equiv \sup_Q \frac{1}{|Q|} \int_Q |f(x) - f_Q|\, dx < \infty. \tag{5.8.1}$$

Here Q ranges over all cubes in $\mathbb{R}^N$ with sides parallel to the axes, and f_Q denotes the average of f over the cube Q; we use the expression $|Q|$ to denote the Lebesgue measure, or volume, of Q. There are a number of equivalent definitions of BMO; we mention two of them:

$$\inf_{c \in \mathbb{C}} \sup_Q \frac{1}{|Q|} \int_Q |f(x) - c|\, dx < \infty \tag{5.8.2}$$

and

$$\sup_Q \left[\frac{1}{|Q|} \int_Q |f(x) - f_Q|^q\, dx \right]^{1/q} < \infty, \quad \text{some } 1 \leq q < \infty. \tag{5.8.3}$$

The latter definition, when $q = 2$, is particularly useful in martingale and probability theory.

It is easy to see that definition (5.8.2) for BMO implies the original definition (5.8.1) of BMO. For

$$\frac{1}{|Q|} \int_Q |f(x) - f_Q|\, dx$$

$$\leq \frac{1}{|Q|} \int_Q |f(x) - c|\, dx + \frac{1}{|Q|} \int_Q |c - f_Q|\, dx$$

$$= \frac{1}{|Q|} \int_Q |f(x) - c|\, dx + \frac{1}{|Q|} \int_Q \left| c - \frac{1}{|Q|} \int_Q f(t)\, dt \right|\, dx$$

$$= \frac{1}{|Q|} \int_Q |f(x) - c|\, dx + \frac{1}{|Q|} \int_Q \left| \frac{1}{|Q|} \int_Q [c - f(t)]\, dt \right|\, dx$$

$$\leq \frac{1}{|Q|} \int_Q |f(x) - c| \, dx + \frac{1}{|Q|} \int_Q \frac{1}{|Q|} \int_Q |c - f(t)| \, dt \, dx$$

$$\leq \frac{1}{|Q|} \int_Q |f(x) - c| \, dx + \frac{1}{|Q|} \int_Q |c - f(t)| \, dt.$$

The converse implication is immediate. Also, definition (5.8.3) of BMO implies the original definition (5.8.1) by an application of Hölder's inequality. The converse implication requires the John-Nirenberg inequality (see the discussion below, as well as[JON]); it is difficult, and we omit it.

It is obvious that $L^\infty \subseteq BMO$. A nontrivial calculation (see [JON]) shows that $\ln |x| \in BMO(\mathbb{R})$. It is noteworthy that

$$(\ln |x|) \cdot \operatorname{sgn} x \notin BMO(\mathbb{R}).$$

We invite the reader to do some calculations to verify these assertions. In particular, accepting these facts we see that the BMO norm is a measure both of size and of smoothness.

Observe that the $\| \ \|_*$ norm is oblivious to constant functions. So the BMO functions are really defined modulo additive constants. Equipped with a quotient norm, BMO is a Banach space. Its first importance for harmonic analysis arose in the following result of Stein: If T is a Calderón-Zygmund operator, then T maps L^∞ to BMO (see [FES] for the history of this result). If we take Stein's result for granted, then we can begin to explore how BMO fits into the infrastructure of harmonic analysis.

To do so, we think of H^1_{Re} in the following way:

$$H^1_{\text{Re}} \ni f \longleftrightarrow (f, R_1 f, R_2 f, \ldots, R_N f) \in (L^1)^{N+1}. \qquad (5.8.4)$$

Suppose that we are interested in calculating the dual of the Banach space H^1_{Re}. Let $\beta \in (H^1_{\text{Re}})^*$. Then, by the Hahn-Banach theorem, there is a continuous extension of β to an element of $[(L^1)^{N+1}]^*$. But we know the dual of L^1 (see Section 0.3), so we know that the extension (which we continue to denote by β) can be represented by integration against an element of $(L^\infty)^{N+1}$. Say that $(g_0, g_1, \ldots, g_N)$ is the representative for β. Then we may calculate, for $f \in H^1_{\text{Re}}$, that

$$\beta(f) = \int_{\mathbb{R}^N} (f, R_1 f, \ldots, R_N f) \cdot (g_0, g_1, \ldots, g_N) \, dx$$

$$= \int_{\mathbb{R}^N} f \cdot g_0 \, dx + \sum_{j=1}^{N} \int_{\mathbb{R}^N} (R_j f) g_j \, dx$$

$$= \int_{\mathbb{R}^N} f \cdot g_0 \, dx - \sum_{j=1}^{N} f(R_j g_j) \, dx.$$

Here we have used the elementary observation, made already in Chapter 0, that the adjoint of a convolution operator with kernel K is the convolution operator with kernel $\widetilde{K}(x) \equiv K(-x)$ (see the proof of Lemma 1.6.15). We finally rewrite the last line as

$$\beta(f) = \int_{\mathbb{R}^N} f \left[g_0 - \sum_{j=1}^{N} R_j g_j \right] dx.$$

But, by the remarks two paragraphs ago, $R_j g_j \in BMO$ for each j. And we have already noted that $L^\infty \subseteq BMO$. As a result, the function

$$g_0 - \sum_{j=1}^{N} R_j g_j \in BMO.$$

We conclude that the dual space of H^1_{Re} has a natural embedding into BMO. Thus we might wonder whether $[H^1_{\text{Re}}]^* = BMO$. The answer to this question is affirmative, and is a deep result of C. Fefferman (see [FES]). We shall not prove it here. However we shall use our understanding of atoms to have a look at the result.

The atomic theory has taught us that a typical H^1_{Re} function is an atom a. Let us verify that any such 1-atom a pairs with a BMO function ϕ. We suppose for simplicity that a is supported in the ball $B = B(0, r)$.

To achieve our goal, we examine

$$\int_{\mathbb{R}^N} a(x)\phi(x) \, dx = \int_B a(x)\phi(x) \, dx.$$

We use the fact that a has mean-value-zero to write this last as

$$\int_B a(x)[\phi(x) - \phi_B] \, dx.$$

Then

$$\left| \int_{\mathbb{R}^N} a(x)\phi(x) \, dx \right| \leq \int_B |a(x)||\phi(x) - \phi_B| \, dx$$

$$\leq \frac{1}{m(B)} \int_B |\phi(x) - \phi_B| \, dx$$

$$\leq \|\phi\|_*.$$

This calculation shows that the BMO function ϕ pairs with any atom, and the bound on the pairing is independent of the particular atom (indeed it depends only on the $\| \ \|_*$ norm of ϕ).

A fundamental fact about BMO functions is the John-Nirenberg inequality (see [JON]). It says, in effect, that a BMO function ϕ has distribution function μ (see Appendix VII) that is comparable to the distribution function of an exponentially integrable function (i.e., a function f such that $e^{c|f|}$ is integrable for some small positive constant c). Here is a more precise statement:

Theorem 5.8.5 (The John-Nirenberg Inequality) *Let f be a function that lies in $BMO(Q_0)$, where Q_0 is a cube lying in $\mathbb{R}^N$ (here we are mandating that f satisfy the BMO condition for subcubes of Q_0 only). Then, for appropriate constants $c_1, c_2 > 0$,*

$$m\{x \in Q_0 : |f(x) - f_{Q_0}| > \lambda\} \leq c_1 e^{-c_2\lambda} |Q_0|. \qquad (5.8.5.1)$$

It follows from (5.8.1.1) that ϕ is in every L^p class (at least locally) for $p < \infty$ (exercise—use Appendix VII). But, as noted at the beginning of this section, BMO functions are not necessarily L^∞.

For many purposes, BMO functions are the correct *ersatz* for L^∞ in the context of harmonic analysis. For instance, it can be shown that any Calderón-Zygmund operator maps BMO to BMO. By duality (since the adjoint of a Calderón-Zygmund operator is also a Calderón-Zygmund operator) it follows that any Calderón-Zygmund operator maps H^1_{Re} to H^1_{Re}. Thus we see that the space H^1_{Re} is a natural substitute for L^1 in the context of harmonic analysis.

The study of real-variable Hardy spaces, and corresponding constructs such as the space BMO of John and Nirenberg, has changed the face of harmonic analysis in the past 25 years. We continue to make new discoveries about these spaces (see, for example [CHKS1], [CHKS2]).

CHAPTER 6

Modern Theories of Integral Operators

6.1 Spaces of Homogeneous Type

One of the important developments in harmonic analysis in the 1960s and 1970s was a realization that, in certain important contexts, an axiomatic theory is possible. What does this assertion mean?

The Brelot potential theory gives a version of axiomatic potential theory [CHO]. That is not the focus of the current discussion. In the classical harmonic analysis of Euclidean space, we are interested in maximal operators (of Hardy-Littlewood type), in fractional integrals (of Riesz type), and in singular integral operators. What structure does a space require in order to support analytic objects like these that enjoy at least a modicum of the familiar regularity theory?

If we wanted to do harmonic analysis on the boundary of a domain in $\mathbb{R}^N$, or on a surface or manifold, or on a Lie group, which of the ideas that we learned in Chapters 1–5 would transfer with ease?

The answer to these questions is obscured by the history of the subject. Classical Fourier analysis is inextricably bound up with the fact that Euclidean space has three natural groups acting on it: translations, dilations, and rotations. The functions $x \mapsto e^{ix\cdot\xi}$, for $\xi \in \mathbb{R}^N$, are the characters of the group $(\mathbb{R}^N, +)$ (see Section 2.1). These are the basic units of harmonic analysis on $\mathbb{R}^N$. They are part and parcel of our study of *translation-invariant* linear operators (also of the so-called *sublinear* operators, such as the Hardy-Littlewood maximal operator). The interaction of the Fourier transform with dilations and with rotations has also strongly influenced the development that we have presented so far.

The works [HOR], [SMI], and most importantly [COIW1] began to show us that many of the most fundamental ideas of harmonic analysis can indeed be developed, from first principles, without any reference to the three classical groups that act on Euclidean space. These new ideas do indeed give us the freedom to do analysis on the boundary of a domain, on a surface, or on a group. In the present sections we shall discuss some of the key ideas that come from [COIW1].

The mathematical structure that we shall develop here is called a "space of homogeneous type." It is quite easy to manipulate and to comprehend at an introductory level. The reader should be cautioned *not* to confuse this new notion with the more classical idea of "homogeneous space" (see [HEL]). There are some connections between the two sets of ideas, but these will be of no interest for us here.

Suppose that we are given a set X that is equipped with a *quasi-metric* ρ. Thus $\rho : X \times X \to \mathbb{R}^+$ satisifies the three axioms

(6.1.1) $\rho(x, y) = 0$ if and only if $x = y$;

(6.1.2) $\rho(x, y) = \rho(y, x)$;

(6.1.3) There is a constant $C_2 > 0$ such that if $x, y, z \in X$, then

$$\rho(x, z) \le C_2 \big[\rho(x, y) + \rho(y, z) \big].$$

We call this the *quasi-triangle inequality*.

We take the balls $B(x, r) = \{ y \in X : \rho(x, y) < r \}$ to be the basis for a topology on X.

Now assume that there is given a measure μ on X. For convenience, we assume that μ is a regular Borel measure, though little will be said about this hypothesis in the sequel (and the reader unfamiliar with the notion should suspend his disbelief). We say that (X, ρ, μ) is a *space of homogeneous type* if the following axioms are satisfied:

(6.1.4) For each $x \in X$ and $r > 0$, $0 < \mu[B(x, r)] < \infty$;

(6.1.5) [The Doubling Property] There is a constant $C_1 > 0$ such that, for any $x \in X$ and $r > 0$, we have

$$\mu[B(x, 2r)] \leq C_1 \cdot \mu[B(x, r)].$$

Example 6.1.6 Let $X = \mathbb{R}^N$, take ρ to be Euclidean distance, and let μ be ordinary Lebesgue measure. The ball $B(x, r)$ will be the usual open Euclidean ball with center x and radius r.

Then it is obvious that each ball has positive, finite measure (Axiom 6.1.4), and that the balls form a basis for the topology of $\mathbb{R}^N$. A change of variable gives that a ball with radius $2r$ has volume which is 2^N times the volume of a ball with radius r (Axiom 6.1.5). $\square$

Example 6.1.7 Let $X = \mathbb{R}^N$, let μ be Lebesgue measure, and define

$$\rho(x, y) = \sum_{j=1}^{N} |x_j - y_j|^{\beta_j},$$

where β_j are positive numbers that are not all equal. This quasi-metric is *nonisotropic* (that is, it measures distance in different coordinate directions differently). It is not difficult to check that ρ satisfies a quasi-triangle inequality, that balls have positive, finite measure, and that the balls satisfy the doubling property. Details are left for the interested reader. $\square$

Example 6.1.8 Let $X = [-1, 1] \subseteq \mathbb{R}^N$. For the measure, take $d\mu(x) = (1 - x)^\alpha (1 + x)^\beta \, dx$, where α, β are real constants that exceed -1. Let ρ be the Euclidean distance. It is straightforward to verify that this setup gives the structure of a space of homogeneous type. It is useful in the study of Jacobi polynomials. $\square$

Example 6.1.9 Let $\Omega \subseteq \mathbb{R}^N$ be a bounded domain with C^2 boundary. Let $X = \partial\Omega$ and let $\mu = \sigma$ be ordinary surface measure on X (i.e., $(N - 1)$-dimensional Hausdorff measure—see Appendix IV). Let the

metric be the ordinary Euclidean distance inherited from $\mathbb{R}^N$. A ball
with center $x \in X$ and radius $r > 0$ is then the intersection of X with
the Euclidean ball of center x and radius r; in other words

$$B(x, r) = \{y \in X : |x - y| < r\}.$$

A moment's thought now shows that each ball has positive, finite
area/volume. One can use the exponential map (see [HEL]—this is the
map that sends each tangent vector to the point at $t = 1$ along the
geodesic $\gamma(t)$ in the direction of that tangent vector) to get explicit
bounds both above and below on the area/volume of a ball. Again using
the exponential map and/or estimates on the curvature of the boundary,
one can see that $\mu(B(x, r)) \approx cr^{N-1}$; hence there is a constant $C > 0$
so that

$$\mu(B(x, 2r)) \leq C \cdot \mu(B(x, r)). \qquad \Box$$

Example 6.1.10 (optional) Let X be a smooth surface in $\mathbb{R}^N$. Let
$Y_1, \ldots, Y_k$ be smooth vector fields on X. Let μ be ordinary surface
measure on X. Say that a C^1 curve $\gamma : [0, r] \rightarrow X$ is *admissible* if, for
each t,

$$\frac{d\gamma}{dt}(t) = \sum_{j=1}^{k} c_j(t) Y_j(\gamma(t)),$$

where $\sum_j |c_j(t)|^2 \leq 1$. If $x, y \in X$, then we set $\rho(x, y)$ equal to
the infimum of the set of all r for which there is an admissible curve
γ with $\gamma(0) = x$ and $\gamma(r) = y$. The verification that ρ satisfies the
quasi-triangle inequality is direct from the definition of the distance ρ.

A standard condition on a family of vector fields, coming from
the study of the Frobenius theorem, or from the theory of hypoelliptic
partial differential operators, is this: that $Y_1, \ldots, Y_k$ together with their
commutators $[Y_j, Y_k] \equiv Y_j Y_k - Y_k Y_j$ and the higher-order commutators $[Y_j, [Y_k, Y_\ell]]$, etc., span all tangent directions at each point. This is
commonly known as the "Hörmander condition." With this condition
in place, it follows that each $B(x, r)$ is an open set in X and that (provided X is connected) any two points can be connected by at least one

admissible curve. It follows from the Hörmander condition that each ball has positive measure. If the vector fields are bounded in magnitude (i.e., $|Y_j(x)| \leq C$ for every j and x), then each $B(x, r)$ will have finite measure. It is actually quite difficult to verify the doubling property (Axiom 6.1.5), and we shall say no more about it here (but see [NSW]).

This particular "space of homogeneous type" structure on a smooth surface is particularly important in the study of hypoellipticity for certain "sum of squares" partial differential operators.   □

Example 6.1.11 Let $B = \{(z_1, z_2) \in \mathbb{C}^2 : |z_1|^2 + |z_2|^2 < 1\}$. This is the unit ball in complex 2-space. We want to equip the boundary $X = \partial B$ with the structure of a space of homogeneous type. For $z, w \in \partial B$, we define

$$\rho(z, w) = \sqrt{|1 - z \cdot \overline{w}|}$$

where, for $z = (z_1, z_2)$, $w = (w_1, w_2)$, $z \cdot \overline{w} \equiv z_1 \overline{w}_1 + z_2 \overline{w}_2$. It turns out (though this is not obvious) that ρ is a genuine metric (see [KRA3, p. 149] for the details). Equipped with ordinary rotationally invariant surface measure (Appendix IV), and the metric and balls just described, ∂B becomes a space of homogeneous type.

Interestingly, the balls constructed in the preceding paragraph are nonisotropic. This means that they do not have the same extent in each direction. Here is a calculation to suggest why this is so. Let us assume for simplicity that $z = (1, 0)$. Then the condition $\rho(z, w) < r$ becomes

$$|1 - w_1| < r^2.$$

Now let us use some information about ∂B. We calculate that

$$|w_2|^2 = 1 - |w_1|^2$$

$$= (1 + |w_1|)(1 - |w_1|)$$

$$\leq 2(1 - |w_1|)$$

$$\leq 2 \cdot |1 - w_1|$$
$$< 2r^2.$$

We conclude that $|w_2| < \sqrt{2}r$.

Our calculations show that the ball centered at $z = (1, 0)$ and with radius r has extent r^2 in the w_1-direction (this is known as the "complex normal direction") and extent $\sqrt{2}r$ in the w_2-direction (this is known as the "complex tangential direction"). The decomposition of space into complex normal and complex tangential directions depends on the point of reference (in this case $(1, 0)$) and changes from point to point of ∂B in a complicated manner. These facts have a profound influence on the harmonic analysis of the unit ball in multidimensional complex space. See [KRA4] for more on these matters, and [KRA3] for information about the metric ρ.   □

Example 6.1.12 For each $x = (x_1, x_2, x_3) \in \mathbb{R}^3$ we define the associated matrix

$$\mathcal{M}_x \equiv \begin{pmatrix} 1 & x_1 & x_3 + \frac{x_1 x_2}{2} \\ 0 & 1 & x_2 \\ 0 & 0 & 1 \end{pmatrix}.$$

Matrix multiplication then gives a (non-abelian) group structure on $\mathbb{R}^3$. In fact, the induced binary operation on $\mathbb{R}^3$ is

$$x \cdot y = \left(x_1 + y_1, x_2 + y_2, x_3 + y_3 + \frac{x_1 y_2 - y_1 x_2}{2} \right).$$

The measure invariant under translation by the group action (i.e., Haar measure) is just ordinary Lebesgue measure.

It is useful on this group (which is a presentation of the Heisenberg group that arises in quantum mechanics) to use the quasi-metric

$$\rho(x, y) = |x_1 - y_1| + |x_2 - y_2| + \left| x_3 - y_3 + \frac{y_1 x_2 - x_1 y_2}{2} \right|^{1/2}.$$

Both the group operation and this notion of distance respect the "dilations" given by $\alpha_\lambda(x) = (\lambda x_1, \lambda x_2, \lambda^2 x_3)$.

It may be checked that $X = \mathbb{R}^3$, equipped with the structures that we have just described, is a space of homogeneous type. This particular space may be identified with the boundary of the unit ball in $\mathbb{C}^2$ in a natural way (see Example 6.1.11), and is useful in studying partial differential equations in that setting (particularly the Lewy unsolvable operator, for which see [KRA3]).   □

The references [CHR] and [COIW2] contain many other examples of spaces of homogeneous type.

Remark. It may be noted that sometimes a space X is presented for which there is a "ball" of radius $r > 0$ assigned to each point $x \in X$, but no metric or quasi-metric defined. Minimal alternative hypotheses are that $B(x, r) \subseteq B(x, s)$ when $r \leq s$, that the balls form a neighborhood basis at each x, and that the resulting topology be that of a Hausdorff space. In these circumstances, one can define

$$\rho(x, y) = \inf\{r > 0 : x, y \in B(z, r) \text{ for some } z \in X\}.$$

If the measure is compatible with the balls in a suitable sense, then a space of homogeneous type results.

6.2   Integral Operators on a Space of Homogeneous Type

Recall that in Section 5.1 we learned about the fractional integral operators of Riesz. These are operators of the form

$$f \mapsto f * R_\alpha,$$

where

$$R_\alpha(x) = |x|^{-N+\alpha}, \quad 0 < \alpha < N.$$

As we saw in the proof of the Sobolev imbedding theorem (Theorem 5.5.4), the Riesz operators serve as fractional-order integration operators (hence the name). The basis for this assertion can be gleaned from an examination of the Fourier transform side.

It is a deep insight from the paper [FOST] that the key properties of the kernels R_α—that force them to have the desired mapping properties—are in fact not Fourier-analytic but instead are measure-theoretic. We now formulate this result:

Theorem 6.2.1 *Suppose that* (X, μ), (Y, ν) *are measure spaces, and that* $K : X \times Y \to \mathbb{C}$ *is a product-measurable function. For each* $y \in Y$ *and each* $\lambda > 0$ *set*

$$E_y(\lambda) = \{x \in X : |K(x, y)| > \lambda\}$$

and for each $x \in X$ *and each* $\lambda > 0$ *set*

$$E^x(\lambda) = \{y \in Y : |K(x, y)| > \lambda\}.$$

Let $1 < r < \infty$. *Assume that there is a constant* $C > 0$ *such that*

$$\mu(E_y(\lambda)) \leq \frac{C}{\lambda^r} \qquad \text{for all } y, \lambda$$

and

$$\nu(E^x(\lambda)) \leq \frac{C}{\lambda^r} \qquad \text{for all } x, \lambda.$$

[*The first inequality says that* $K(\,\cdot\,, y)$ *is weak-type* r, *uniformly in the parameter* y. *The second inequality says that* $K(x, \,\cdot\,)$ *is weak-type* r, *uniformly in the parameter* x.] *Then the operator*

$$T : f \longmapsto \int_Y f(y) K(x, y) \, d\nu(y),$$

which is initially defined for simple functions f, *extends to be a bounded operator from* L^p *to* L^q, *where* $1 < p < r' \equiv r/(r-1)$

and

$$\frac{1}{q} = \frac{1}{p} + \frac{1}{r} - 1.$$

Proof. The idea of the proof is one that is pervasive in the subject: First, we fix the constant $\lambda > 0$. Then we *divide* the kernel at a place that depends on the constant λ. Finally, we apply Schur's lemma and the Chebyshev Inequality to each of those pieces. Here are the details, which closely follow the argument in [FOST].

Fix $1 < p < r$. By the Marcinkiewicz interpolation theorem (Appendix III), it is enough for us to show that T is weak-type (p, q), where $1/q = 1/p + 1/r - 1$ and $1 \le p < q < \infty$. Fix a number $\lambda > 0$ (we shall specify λ more precisely below). We begin by defining

$$K_1(x, y) = \begin{cases} K(x, y) & \text{if } |K(x, y)| \ge \lambda \\ 0 & \text{if } |K(x, y)| < \lambda \end{cases}$$

and

$$K_2(x, y) = \begin{cases} 0 & \text{if } |K(x, y)| \ge \lambda \\ K(x, y) & \text{if } |K(x, y)| < \lambda. \end{cases}$$

For each simple function f we define

$$T_1 f(x) = \int_Y f(y) K_1(x, y) \, d\nu(y)$$

and

$$T_2 f(x) = \int_Y f(y) K_2(x, y) \, d\nu(y).$$

Clearly, $T = T_1 + T_2$. Hence, for any $\sigma > 0$,

$$\mu\{x : |Tf(x)| > 2\sigma\} \le \mu\{x : |T_1 f(x)| > \sigma\} + \mu\{x : |T_2 f(x)| > \sigma\}.$$

So it is enough for us to estimate the two expressions on the right.

Now fix a simple function f. We may as well suppose that $\|f\|_{L^p} = 1$. Let p' be such that $1/p + 1/p' = 1$. Hölder's inequality tells us that, for every $x \in X$,

$$|T_2 f(x)| \leq \left(\int |K_2(x, y)|^{p'} d\nu(y) \right)^{1/p'} \cdot \|f\|_{L^p}$$

$$= \left(\int |K_2(x, y)|^{p'} d\nu(y) \right)^{1/p'}.$$

For a function g on Y, set $\mu_g(s) = \nu(\{y \in Y : |g(y)| > s\})$. Then (using Appendix VII)

$$\int |K_2(x, y)|^{p'} d\nu(y) = p' \int_0^\lambda s^{p'-1} \mu_{K(x, \cdot)}(s) \, ds$$

$$\leq Cp' \int_0^\lambda s^{p'-1-r} \, ds$$

$$\leq C' \lambda^{p'-r}.$$

[Note that $1 < p < r'$ so that $p' > r$ and $p' - 1 - r > -1$, which is essential for the convergence of the integral.] Since $1 - r/p' = r/q$, we see that $|T_2 f(x)| \leq C_0 \lambda^{1-(r/p')} = C_0 \lambda^{r/q}$. For any number $\sigma > 0$, we set $\lambda = \lambda(\sigma) = (\sigma/C_0)^{q/r}$. Thus $\mu\{x : |T_2 f(x)| > \sigma\} = 0$.

Our next observation is that, since $r > 1$ (and using Appendix VII),

$$\int |K_1(x, y)| \, d\nu(y) = \int_\lambda^\infty \mu_{K(s, \cdot)}(s) \, ds \leq C \int_\lambda^\infty s^{-r} \, ds \leq C \lambda^{1-r}.$$

A similar argument shows that $\int_X |K_1(x, y)| \, d\mu(x) \leq C \lambda^{1-r}$ for all[1] $y \in Y$. Then Schur's lemma (Lemma 1.4.5) shows that T_1 is bounded on L^p and that

$$\|T_1 f\|_{L^p} \leq C \lambda^{1-r} \|f\|_{L^p} = C \lambda^{1-r}.$$

---

[1]As usual, we indulge in the custom of using the single character C to denote different constants in different occurrences.

Therefore, by Chebyshev's inequality (Lemma 0.2.17),

$$\mu\{x : |T_1 f(x)| > \sigma\} \leq \frac{\|T_1 f\|_{L^p}^p}{\sigma^p}$$

$$\leq C(\lambda^{1-r}/\sigma)^p$$

$$= C\sigma^{[(1-r)qp/r]-p}$$

$$\leq C\sigma^{-q}$$

$$= C(\|f\|_{L^p}/\sigma)^q.$$

This completes the proof. □

Notice that our treatment of fractional integral operators in Theorem 6.2.1 is certainly valid on a space of homogeneous type—indeed on any measure space. And it avoids the notion of homogeneity of the kernel. The classical integral kernel R_α for the fractional integral operator I_α (see Section 5.1) is homogeneous of degree $-N + \alpha$, and that fact played a role in the classical proofs (see [STE1]). But now we can instead consider the weak-type of the kernel:

$$m\{x \in \mathbb{R}^N : |x|^{-N+\alpha} > \sigma\} = m\{x \in \mathbb{R}^N : |x|^{N-\alpha} < 1/\sigma\}$$

$$= m\{x \in \mathbb{R}^N : |x| < \sigma^{-1/(N-\alpha)}\}$$

$$= m(B(0, \sigma^{-1/(N-\alpha)}))$$

$$= \sigma^{-N/[N-\alpha]} \cdot m(B(0, 1)).$$

Thus we see that the kernel k_α is of weak-type $N/[N - \alpha]$; hence $k_\alpha(x - y)$ is uniformly of weak-type $N/(N - \alpha)$ in x and in y. By the theorem, convolution with this kernel maps L^p to L^q, where

$$\frac{1}{q} = \frac{1}{p} + \frac{1}{N/[N - \alpha]} - 1 = \frac{1}{p} - \frac{\alpha}{N}.$$

This is the result that we described in Theorem 5.1.3. Now we have proved it completely.

Likewise, our treatment below of the other basic ingredient of harmonic analysis—singular integrals—will also avoid the classical issue of homogeneity. When we discussed singular integrals in Chapter 5, we enunciated but did not prove the fundamental Calderón-Zygmund theorem. Now we shall flesh out the theory in two ways: **(i)** We shall generalize the notion of singular integral operator to a space of homogeneous type and **(ii)** we shall provide the details of the proof of the Calderón-Zygmund theorem in this more general setting.

We assume to begin that we are working on a space X of homogeneous type with measure μ. We assume that Axioms (6.1.4)–(6.1.5) are in place. We begin with that most fundamental of objects in harmonic analysis, the Hardy-Littlewood maximal function. If f is a locally integrable function on (X, μ), then we set

$$Mf(x) = \sup_{r>0} \frac{1}{\mu(B(x,r))} \int_{B(x,r)} |f(t)| \, d\mu(t).$$

We need to know that M has certain mapping properties, and for this we need a covering lemma. Note as an exercise that Mf is measurable when f is, because in fact Mf is lower semicontinuous.

Lemma 6.2.2 *Let $K \subseteq X$ be a compact set. Let $\{B(x_\alpha, r_\alpha)\}$ be a collection of open balls that covers K. Then there is a subcollection $\{B(x_{\alpha_j}, r_{\alpha_j})\}$ that is pairwise disjoint and such that $\{B(x_{\alpha_j}, C_3 r_{\alpha_j})\}$ still covers K, where $C_3 = 2C_2^2 + C_2$.*

Note here that the constant C_2 in the statement of the lemma is the same constant that occurred in the quasi-triangle inequality (6.1.3) for the quasi-metric ρ.

Proof. [This argument is quite similar to the proof of Lemma 1.5.2.] Since K is compact, we may suppose at the outset that we are dealing with a finite collection of balls $B(x_k, r_k)$. Now select a ball $B(x_{k_1}, r_{k_1})$ of greatest radius (if two or more have the same greatest radius, then just pick one). Now select $B(x_{k_2}, r_{k_2})$ from among those balls remaining to **(i)** have greatest possible radius, and **(ii)** be disjoint from $B(x_{k_1}, r_{k_1})$.

Continue in this fashion: At the mth step, select a (not previously selected) ball $B(x_{k_m}, r_{k_m})$ that **(i)** has greatest possible radius and **(ii)** is disjoint from each of the $(m - 1)$ preceding balls $B(x_{k_1}, r_{k_1}), \ldots, B(x_{k_{m-1}}, r_{k_{m-1}})$. Since there are only finitely many balls in the entire collection, the process must eventually stop. We claim that this collection of balls does the job.

The designated collection of balls is certainly pairwise disjoint by the way that we selected each ball in succession. Now we need to see that $\{B(x_{k_\ell}, C_3 r_{k_\ell})\}$ covers K (of course $C_3 \geq 1$). It suffices to see that these dilated balls cover the original collection $B(x_k, r_k)$, for these latter balls were assumed to cover K. Now fix attention on one of the original balls $B(x_k, r_k)$. If it is one of the selected balls $B(x_{k_\ell}, r_{k_\ell})$, then it is covered by $\{B(x_{k_\ell}, C_3 r_{k_\ell})\}$. If it is *not* one of the selected balls, then there is a first selected ball $B(x_{k_p}, r_{k_p})$ that intersects it (if there were no such selected ball, then we would have ceased the selection process too soon). Remember that the selected balls are ordered according to decreasing size! By the selection scheme, it must be that $r_{k_p} \geq r_k$, otherwise we chose $\{B(x_{k_p}, r_{k_p})\}$ incorrectly.

Now we claim that $B(x_k, r_k) \subseteq B(x_{k_p}, C_3 r_{k_p})$. To see this, let $z \in B(x_k, r_k) \cap B(x_{k_p}, r_{k_p})$ and let $x \in B(x_k, r_k)$ be arbitrary. Then

$$
\begin{aligned}
\rho(x, x_{k_p}) &\leq C_2\big[\rho(x, z) + \rho(z, x_{k_p})\big] \\
&\leq C_2\big[C_2(\rho(x, x_k) + \rho(x_k, z)) + r_{k_p}\big] \\
&\leq C_2\big[C_2(r_k + r_k) + r_{k_p}\big] \\
&\leq r_{k_p}\big[2C_2^2 + C_2\big].
\end{aligned}
$$

As a result, $B(x_k, r_k) \subseteq B(x_{k_p}, C_3 r_{k_p})$. That is what we wanted to know. $\square$

Remark. We took care to state the last lemma for a covering of a compact set. And the property of compactness was convenient in the proof, for it guaranteed that the selection process would stop. However, the lemma is still true when the set that is being covered is open. The additional ideas needed can be found in [COIW1, p. 69 ff.]. In the applications that we shall see, one can write the open set in question as a union of compact sets K_j, with each K_j the closure of an open set and $\overset{\circ}{K}_j \cap \overset{\circ}{K}_\ell = \emptyset$ when $|j - \ell| > 2$. Then apply 6.2.2 as stated to each K_j; any duplicate balls that come from adjacent K_j and K_ℓ can then be discarded as long as we replace the constant C_3 in the conclusion of the lemma by $2(C_3 + 2)$.

Proposition 6.2.3 *The Hardy-Littlewood maximal operator is weak-type* $(1, 1)$.

Proof. [This proof is similar to that of 1.5.3.] Fix a function $f \in L^1$. Fix a constant $\lambda > 0$. By the inner regularity of the measure μ, it is enough to estimate $\mu(K)$, where K is any compact subset of $\{x \in X : Mf(x) > \lambda\}$. If $x \in K$ then, by definition, there is a number $r_x > 0$ such that

$$\frac{1}{\mu(B(x, r_x))} \int_{B(x,r_x)} |f(t)| \, d\mu(t) > \lambda.$$

It is convenient to rewrite this inequality as

$$\mu(B(x, r_x)) < \frac{1}{\lambda} \int_{B(x,r_x)} |f(t)| \, d\mu(t).$$

Now the balls $B(x, r_x)$ certainly cover K. By the lemma, we may choose a disjoint subcollection $\{B(x_k, r_{x_k})\}$ whose C_3-fold dilates cover K. Then (letting $[s]$ denote the greatest integer in s)

$$\mu(K) \leq \mu \left(\cup_k B(x_k, C_3 r_{x_k}) \right)$$
$$\leq \sum_k \mu(B(x_k, C_3 r_{x_k}))$$
$$\leq \sum_k C_1^{[\log_2 C_3]+1} \mu(B(x_k, r_{x_k}))$$

[Notice here that we are repeatedly using Axiom 6.1.5.]

$$\leq \sum_k C_1^{[\log_2 C_3]+1} \frac{1}{\lambda} \int_{B(x,r_{x_k})} |f(t)| \, d\mu(t)$$

$$\leq \frac{C_1^{[\log_2 C_3]+1}}{\lambda} \int_X |f(t)| \, d\mu(t)$$

$$\equiv \frac{C'}{\lambda} \|f\|_{L^1}.$$

This proves the assertion. □

Corollary of the Proof: *Define the modified Hardy-Littlewood maximal operator*

$$M^* f(x) = \sup_{B \ni x} \frac{1}{\mu(B)} \int_B |f(t)| \, d\mu(t),$$

where B ranges over balls that contain x. Then M^* is weak-type $(1, 1)$.
We leave the details of the proof as an exercise.

Of course the Hardy-Littlewood maximal operator is trivially bounded on L^∞. By the Marcinkiewicz interpolation theorem, as discussed in Appendix III, we may conclude from Proposition 6.2.3 that the Hardy-Littlewood maximal operator is bounded on L^p for $1 < p < \infty$.

An immediate consequence of the weak-type $(1, 1)$ property of the Hardy-Littlewood operator, and one that we will call upon later, is the Lebesgue differentiation theorem (refer to Lemma 0.2.15):

Theorem 6.2.4 *Let f be a locally L^1 function on X. Then for μ-almost every point $x \in X$ we have*

$$\lim_{r \to 0^+} \frac{1}{\mu(B(x,r))} \int_{B(x,r)} f(t) \, d\mu(t) = f(x).$$

Proof. This result is proved in a manner parallel to, but simpler than, the proof of pointwise convergence for the Poisson summation method in single-variable Fourier series (Section 1.5). We leave the details for the interested reader.   □

As in the classical setting, there is the following stronger form of the Lebesgue differentiation theorem:

Theorem 6.2.5 *Let f be a locally L^1 function on X. Then, for μ-almost every point $x \in X$,*

$$\lim_{r \to 0^+} \frac{1}{\mu(B(x,r))} \int_{B(x,r)} |f(t) - f(x)| \, d\mu(t) = 0.$$

Now we turn to the fundamental lemma of the Calderón-Zygmund theory. As a preliminary result, we need a decomposition of open sets that is analogous to the classical Whitney decomposition (Figure 1). Our treatment stems from that in [COIW1]. The Whitney decomposition is analogous to the classical result that an open set in $\mathbb{R}$ is the pairwise disjoint union of open intervals. Instead, on $\mathbb{R}^N$ with $N > 1$, we decompose a given open set $\mathcal{O}$ into closed cubes with pairwise disjoint interiors, such that the diameter of each cube is comparable to its distance to the boundary of $\mathcal{O}$. On a general space of homogeneous type, we cannot achieve the precision that is given by disjoint cubes, but the reader will see that the spirit of the original is retained (see Lemma 6.2.7 below).

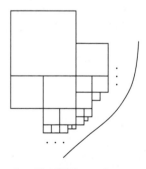

Figure 1. The Whitney decomposition.

Sublemma 6.2.6 *There is a universal constant* K *so that if* $r > 0$ *and if* $B(x_1, r_1)$, $B(x_2, r_2)$, ..., $B(x_p, r_p)$ *are pairwise disjoint balls all having radii exceeding* $r/2$ *and all contained in* $B(y, r)$ *for some point* y, *then* $p \leq K$.

Proof. Let ρ be the quasi-metric introduced in Section 6.1. Clearly $\rho(y, x_i) < r$ for each i. If z is any point of $B(y, r)$ and $1 \leq i \leq p$ then, by the pseudo-triangle inequality,

$$\rho(x_i, z) \leq C_2\left[\rho(x_i, y) + \rho(y, z)\right]$$
$$< C_2 \cdot 2r.$$

It follows that $B(y, r) \subseteq B(x_i, 2C_2 r)$ for any i. As a result

$$\sum_j \mu(B(x_j, r/2)) \leq \sum_j \mu(B(x_j, r_j)) \leq \mu(B(y, r))$$
$$\leq \mu(B(x_i, 2C_2 r))$$

for any $1 \leq i \leq p$. Using Axiom 6.1.5 repeatedly, we may majorize the right side by $K\mu(B(x_i, r/2))$, some positive constant K. Summing over i yields

$$p \cdot \sum_j \mu(B(x_j, r/2)) \leq K \sum_i \mu\big(B(x_i, r/2)\big).$$

As a result, $p \leq K$. □

Remark. Of course obvious variants of Sublemma 6.2.6 are also true. For any integer $M > 0$, there is an *a priori* upper bound K_M such that not more than K_M pairwise disjoint balls of radius r/M can lie in $B(y, r)$.

In practice, the actual value of K is of no interest. However, the reader may wish, as an exercise, to estimate the size of this constant when the space of homogeneous type is $\mathbb{R}^N$. [In the language of [FE1], Sublemma 6.2.6 says that X is a "directionally limited" metric space.]

Lemma 6.2.7 (The Whitney Decomposition) *There are constants* h, $K' > 0$ *with the following property: Let* $\mathcal{O} \subseteq X$ *be a bounded open set (i.e., the set is contained in a ball) with* $\mathcal{O} \neq X$. *Then there is a sequence of balls* $B(o_j, r_j)$ *such that*

(6.2.7.1) $\mathcal{O} = \cup_j B(o_j, r_j)$;

(6.2.7.2) *No point of* $\mathcal{O}$ *belongs to more than* K' *of the balls;*

(6.2.7.3) *Each ball* $B(o_j, hr_j)$ *intersects* $X \setminus \mathcal{O}$.

Proof. For each point $o \in \mathcal{O}$ we consider the ball

$$B_o \equiv B(o, \delta_o/[2 \cdot C_3 \cdot M]),$$

where $M = C_2 + 2$, C_2 is from the quasi-triangle inequality, C_3 is from Lemma 6.2.2, and

$$\delta_o = \rho(o, X \setminus \mathcal{O}) \equiv \inf\{\rho(o, p) : p \in X \setminus \mathcal{O}\}.$$

The balls $\{B_o\}$ certainly cover $\mathcal{O}$, and we may apply Lemma 6.2.2 and the remark following it to find a pairwise disjoint subcollection $B(o_j, \rho_j)$, $\rho_j \equiv \delta_{o_j}/[2 \cdot C_3 \cdot M]$, such that $\{B(o_j, C_2\rho_j)\}$ cover $\mathcal{O}$.

We claim that the balls $B(o_j, r_j) \equiv B(o_j, C_3\rho_j)$ satisfy the conclusions of this lemma.

Certainly conclusion (6.2.7.1) follows from our construction. In addition, each $B(o_j, 2(M+1)r_j)$ intersects $X \setminus \mathcal{O}$. So conclusion (6.2.7.3) is satisfied with $h = 2(M+1)$. It remains to verify conclusion (6.2.7.2).

Suppose that there is some point $y \in \mathcal{O}$ that lies in K' of the balls $B(o_j, r_j)$. Then each of the balls $B(o_j, r_j)$ lies in $B(y, 2C_2r_j)$. But notice that

$$\delta_{o_j} \leq C_2[\delta_y + \rho(y, o_j)] \leq C_2[\delta_y + r_j];$$

hence

$$\delta_y \geq \frac{\delta_{o_j}}{C_2} - r_j$$

$$= \frac{\delta_{o_j}}{C_2} - \frac{\delta_{o_j}}{2M}$$

$$\geq \frac{\delta_{o_j}}{2M}$$

$$= r_j.$$

We conclude that $B(y, 2C_2 r_j) \subseteq B(y, 2C_2 \delta_y)$, so that $B(o_j, r_j) \subseteq B(y, 2C_2 \delta_y)$.

But then each of the pairwise disjoint balls $B(o_j, \rho_j)$ lies in the fixed ball $B(y, 2C_2 \delta_y)$. Finally, for each j,

$$\rho(y, X \setminus \mathcal{O}) \leq C_2 [\rho(y, o_j) + \rho(o_j, X \setminus \mathcal{O})]$$

$$\leq C_2 [r_j + 2(M + 1)r_j]$$

$$\leq M[3 + 2M]r_j.$$

It follows that

$$r_j \geq \frac{\delta_y}{M(3 + 2M)}.$$

Thus the ball $B(y, 2C_2 \delta_y)$ contains a total of K' pairwise disjoint balls each having radius at least $\delta_y / M(3 + 2M)$. This is a contradiction of Sublemma 6.2.6 (and the remark following it) if K' is large enough.

$\square$

Lemma 6.2.8 (Calderón-Zygmund) *Let (X, μ) be a space of homogeneous type. Let f be a nonnegative L^1 function on X, and assume that f has bounded support—i.e., that the support of f lies in some ball B. Fix a number $\alpha > \|f\|_{L^1}/\mu(B)$. Then there exists a sequence of balls $B(x_i, r_i)$ and a constant $C > 0$ such that*

(6.2.8.1) $f(x) \leq C \cdot \alpha$ *for almost every $x \in X \setminus \bigcup_i B(x_i, r_i)$;*

(6.2.8.2) $\dfrac{1}{\mu(B(x_i, r_i))} \displaystyle\int_{B(x_i, r_i)} f(t) \, d\mu(t) \leq C \cdot \alpha$;

(6.2.8.3) $\displaystyle\sum_{i=1}^{\infty} \mu(B(x_i, r_i)) \leq \frac{C}{\alpha} \int_X f(t)\, d\mu(t);$

(6.2.8.4) *There is a constant $K' > 0$ such that no point of X belongs to more than K' of the balls $B(x_i, r_i)$.*

Proof. Fix α and f as in the statement of the lemma, and let C be the weak-type bound derived for the operator M^* in Lemma 6.2.3 and the subsequent discussion.[2] Let $\mathcal{O}_\alpha \equiv \{x \in X : M^* f(x) > C\alpha\}$. By the semicontinuity of the maximal function, the set $\mathcal{O}_\alpha$ is plainly open.

For simplicity, let us suppose at first that X is bounded. We apply the Whitney decomposition lemma to write

$$\mathcal{O}_\alpha = \bigcup_j B(x_j, r_j).$$

[Note that if either C or α is large enough, then $\mathcal{O}_\alpha \neq X$, as is required by the Whitney decomposition lemma.]

Observe that conclusion (6.2.8.1) holds because $f \leq Mf$ almost everywhere on $X \setminus \mathcal{O}_\alpha$ (by the Lebesgue differentiation theorem, 0.2.15 and 6.2.4); and $Mf \leq M^* f \leq C\alpha$ off $\mathcal{O}_\alpha$.

For conclusion (6.2.8.2), consider one of the x_j. Select a point $y_j \in X \setminus \mathcal{O}_\alpha$ such that $y_j \in B(x_j, hr_j)$. It follows that

$$\frac{1}{\mu(B(x_j, hr_j))} \int_{B(x_j, hr_j)} f(t) d\mu(t) \leq M^* f(y_j) \leq C\alpha.$$

But then

$$\frac{1}{\mu(B(x_j, r_j))} \int_{B(x_j, r_j)} f(t) d\mu(t) \leq \frac{C}{\mu(B(x_j, hr_j))} \int_{B(x_j, r_j)} f(t) d\mu(t)$$

(by Axiom 6.1.5, since $h > 1$), and this is clearly

$$\leq \frac{1}{\mu(B(x_j, hr_j))} \int_{B(x_j, hr_j)} f(t) d\mu(t) \leq C\alpha.$$

---

[2]The operator M^* is the modified Hardy-Littlewood maximal operator that we defined in the corollary to the proof of Proposition 6.2.3.

Of course conclusion (6.2.8.3) is the weak-type estimate for the operator M^*:

$$\sum_i \mu(B(x_i, r_i)) \leq K' \mu(\mathcal{O}_\alpha) \leq \frac{C}{\alpha} \int_X f(t) \, d\mu(t),$$

where we have used the fact that no point of $\mathcal{O}_\alpha$ lies in more than K' of the balls $B(x_i, r_i)$.

Conclusion (6.2.8.4) has just been noted.

The case of X unbounded follows by exhausting X by bounded subregions. $\qquad\square$

The standard application of the Calderón-Zygmund lemma is to produce a certain decomposition of a nonnegative L^1 function f. This we now establish.

Proposition 6.2.9 *Let f, α, K', C be as in the preceding lemma. Then there exists a function g (the "good part" of f) and functions h_j (the "bad" pieces of f) with the following properties:*

(6.2.9.1) $f = g + \sum_j h_j$;

(6.2.9.2) $\forall x \in X, |g(x)| \leq C\alpha$;

(6.2.9.3) $\|g\|_{L^1} \leq K' \|f\|_{L^1}$;

(6.2.9.4) $\forall j$, h_j is supported in a ball $B(x_j, r_j)$;

(6.2.9.5) $\forall j$, $\displaystyle\int h_j(t) \, d\mu(t) = 0$;

(6.2.9.6) $\displaystyle\sum_j \|h_j\|_{L^1} \leq 2K' \|f\|_{L^1}$.

Proof. We use the result of the preceding lemma.

Let χ_j be the characteristic function of the ball $B(x_j, r_j)$. Define

$$\psi_j(x) = \begin{cases} \dfrac{\chi_j(x)}{\sum_\ell \chi_\ell(x)} & \text{if} \quad x \in B(x_j, r_j) \\[2ex] 0 & \text{if} \quad x \notin B(x_j, r_j). \end{cases}$$

Notice that, for any fixed value of $x \in \cup_i B(x_i, r_i) = \mathcal{O}_\alpha$, the sum in the denominator is at least 1, and it is at most K' by the K'-condition on the valence of the covering. Further, it is clear that $\sum_j \psi_j$ is the characteristic function of $\mathcal{O}_\alpha$. Set

$$g(x) = \begin{cases} f(x) & \text{if } x \notin \mathcal{O}_\alpha \\ \sum_j \left(\frac{1}{\mu(B(x_j, r_j))} \int_{B(x_j, r_j)} f(t) \psi_j(t) \, d\mu(t) \right) \chi_j(x) & \text{if } x \in \mathcal{O}_\alpha. \end{cases}$$

Also define

$$h_j = f \cdot \psi_j - \left(\frac{1}{\mu(B(x_j, r_j))} \int_{B(x_j, r_j)} f(t) \psi_j(t) \, d\mu(t) \right) \chi_j.$$

Now the desired properties (6.2.9.1)–(6.2.9.6) are clear from the preceding lemma. $\qquad \square$

At long last, we formulate and prove our theorem about singular integral operators. Observe that the statement of the theorem bears only scant resemblance to the very classical version of the Calderón-Zygmund theorem that we discussed in Section 5.2. Recall that, in the classical formulation, the notion of a singular integral kernel relied on "homogeneity" and a mean-value condition. It is not at all clear how to formulate either of these concepts in the more general setting of a space of homogeneous type. Of course we will ultimately give an explicit description of the relation between classical singular integrals of Calderón-Zygmund type and the singular integrals discussed in Theorem 6.2.10.

As our discussion of fractional integration (Theorems 6.2.1, 5.1.3) indicates, the mapping properties of an integral kernel can often be discerned from measure-theoretic properties of the kernel rather than properties that are closely linked to the structure of Euclidean space. Our present approach to singular integrals (see [COIW1] for the origin of these ideas) bears out this new philosophy.

Theorem 6.2.10 Let $k(x, y)$ be a jointly measurable function that lies in $L^2(X \times X, \mu \otimes \mu)$. Assume that the operator that is formally defined (*for* f *a simple function*) *by*

$$T_k f(x) \equiv \int_X k(x, y) f(y) \, d\mu(y)$$

satisfies

(6.2.10.1) *There is a constant* M_1 *such that* $\|T_k f\|_{L^2} \le M_1 \|f\|_{L^2}$;

(6.2.10.2) *There are constants* M_2, M_3 *such that*

$$\int_{\rho(x,y) \ge M_2 \rho(y,y')} |k(x, y) - k(x, y')| \, d\mu(x) < M_3 \qquad \forall y, y' \in X.$$

[*Condition 2 of this theorem is sometimes called the "Hörmander condition."*]

Then, for all $1 < p \le 2$, *there is a constant* $A_p = A_p(M_1, M_2, M_3)$ *such that*

$$\|T_k f\|_{L^p} \le A_p \|f\|_{L^p} \qquad \forall f \in L^2 \cap L^p.$$

Furthermore, when $p = 1$, *there is a constant* A_1 *such that*

$$\mu\{x : |T_k f(x)| > \alpha\} \le A_1 \frac{\|f\|_{L^1}}{\alpha} \qquad \forall f \in L^1.$$

Remark. Before we begin the proof, we should make some explanatory remarks about the form of this theorem and the context into which it fits.

The paradigm for our idea of what a singular integral kernel is comes from the Hilbert transform kernel $1/t$, and more generally from Calderón-Zygmund singular integral kernels. *These kernels do not satisfy the hypothesis* $k \in L^2(X \times X, \mu \otimes \mu)$ *of Theorem 6.2.10.*

The rationale is this: We will apply Theorem 6.2.10 not to a standard Calderón-Zygmund kernel K, but rather to its truncation, for

$N > 1$, given by

$$K_N(x) \equiv \begin{cases} K(x) & \text{if} \quad 1/N < |x| < N \\ 0 & \text{if} \quad |x| \leq 1/N \text{ or } |x| \geq N. \end{cases}$$

Certainly K_N is in $L^2(\mathbb{R}^N \times \mathbb{R}^N)$, since it is bounded and has compact support.

The spirit of Theorem 6.2.10 is that it gives us a bound on the operator norm of the operator T_N induced by K_N, and that bound can be taken to be independent of N. Then we can hope to pass to the original operator, that induced by K, by applying Functional Analysis Principle I while letting $N \to \infty$.

Theorem 6.2.10 of Coifman and Weiss points up the fact that, for an operator with kernel satisfying the Hörmander condition, once you have a good L^2 theory then you also get a good L^p theory. It was in part for this reason that people concentrated on conditions that would imply an L^2 theorem. These investigations tie in nicely with the Cotlar-Knapp-Stein theorem (Theorem 6.4.3 below) and were also part of the motivation for the $T(\mathbf{1})$ theorem (Theorem 6.4.6 below).

Proof of the Theorem. We assume for simplicity that X is bounded. Thus we need only prove the weak-type $(1, 1)$ inequality for α sufficiently large. Once we have that, then we may use the Marcinkiewicz interpolation theorem to obtain the desired L^p estimates, $1 < p < 2$. Thus we will assume that $\alpha > \|f\|_{L^1}/\mu(X)$. By linearity we may assume that $f \geq 0$. And, since X is bounded, the support of f is bounded.

Now let α be as above. We write

$$f = g + \sum_j h_j$$

as in the preceding proposition. The triangle inequality now implies that

$$\mu\{x : |T_k f(x)| > \alpha\} \leq \mu\left(\left\{x : |T_k g(x)| > \frac{\alpha}{2}\right\}\right)$$
$$+ \mu\left(\left\{x : \sum_j |T_k h_j(x)| > \frac{\alpha}{2}\right\}\right).$$

Now observe, first by Chebyshev's inequality and then by the L^2 boundedness of T_k, that

$$\mu\left(\left\{x : |T_kg(x)| > \frac{\alpha}{2}\right\}\right) \leq \frac{4}{\alpha^2}\int |T_kg|^2\,d\mu$$

$$\leq \frac{4M_1^2}{\alpha^2}\|g\|_2^2$$

$$\leq \frac{4M_1^2}{\alpha^2}\cdot C\alpha \cdot \int |g|\,d\mu \qquad \text{(by 6.2.9.2)}$$

$$\leq 4C\cdot K' \cdot \frac{M_1^2}{\alpha}\|f\|_{L^1}. \qquad \text{(by 6.2.9.3)}$$

This is a favorable estimate for T_kg.

Set

$$S \equiv \left\{x : \sum_j |T_kh_j(x)| > \frac{\alpha}{2}\right\}.$$

Use Lemma 6.2.8 to obtain x_j and r_j such that the set

$$S = \left(S \cap \bigcup_j B(x_j, M_2r_j)\right) \bigcup \left(S \setminus \bigcup_j B(x_j, M_2r_j)\right) \equiv S_1 \cup S_2$$

satisfies

$$\mu(S_1) \leq \mu\left(\bigcup_j B(x_j, M_2r_j)\right) \leq \frac{C}{\alpha}\|f\|_{L^1}.$$

It remains to estimate the measure of S_2.

Now Chebyshev's inequality implies that

$$\mu\left\{x \in X \setminus \bigcup_j B(x_j, M_2r_j) : \sum_\ell |T_kh_\ell(x)| > \alpha/2\right\}$$

$$\leq \frac{1}{\alpha/2}\int_{X\setminus\bigcup_j B(x_j, M_2r_j)} \sum_\ell |T_kh_\ell|\,d\mu$$

$$= \frac{2}{\alpha}\sum_\ell \int_{X\setminus\bigcup_j B(x_j, M_2r_j)} |T_kh_\ell|\,d\mu$$

$$\leq \frac{2}{\alpha} \sum_j \int_{X \setminus B(x_j, M_2 r_j)} \left| T_k h_j \right| d\mu$$

$$= \frac{2}{\alpha} \sum_j \int_{\{y:\rho(y,x_j)>M_2 r_j\}} \left| \int k(y,x) h_j(x) \, d\mu(x) \right| d\mu(y)$$

$$= \frac{2}{\alpha} \sum_j \int_{\{y:\rho(y,x_j)>M_2 r_j\}} \left| \int [k(y,x) \right.$$

$$\left. - k(y,x_j)] \, h_j(x) \, d\mu(x) \right| d\mu(y)$$

because $\int h_j(x) \, d\mu(x) = 0$. This last expression is majorized by

$$\frac{C}{\alpha} \sum_j \int \left[\int_{\{y:\rho(y,x_j)>M_2 \rho(x,x_j)\}} \left| k(y,x) \right. \right.$$

$$\left. \left. - k(y,x_j) \right| d\mu(y) \right] |h_j(x)| d\mu(x)$$

because the support of h_j lies in the ball $B(x_j, r_j)$. But of course the y integral is assumed to be bounded by M_3 and the sum of the x-integrals of the $|h_j|$ is known by (6.2.9.6) to be bounded by $2K' \|f\|_{L^1}$. That completes our estimates. $\qquad \square$

Note again that once we have both weak-type $(1, 1)$ estimates and strong-type L^2 estimates for the linear operator T_k, then the L^p estimates follow automatically from the Marcinkiewicz interpolation theorem (Appendix III). The derivation of L^p estimates for $2 < p < \infty$ would require some additional properties of the kernel k—such as some symmetry in the x and y variables. We shall omit a discussion of those estimates.

We conclude this long (and difficult) section by indicating the relationship between the "Hörmander condition" defined in the Calderón-Zygmund theorem and the more classical homogeneity and mean-value-zero condition that we formulated in Section 5.2. For the sake of this discussion, let us suppose that we are back on $\mathbb{R}^N$. Let

$$K(x) = \frac{\Omega(x)}{|x|^N}$$

be a standard Calderón-Zygmund kernel of the classical type. In particular, we assume that Ω is homogeneous of degree zero, has mean-value 0 on the unit sphere, and is smooth away from the origin. Using the standard Euclidean notion of distance to formulate our metric and hence our balls, we calculate that

$$\int_{\{x:|x-y|>2|y-y'|\}} |K(x-y) - K(x-y')| \, dx$$

$$= \int_{\{x:|x-y|>2|y-y'|\}} \left| \frac{\Omega(x-y)}{|x-y|^N} - \frac{\Omega(x-y')}{|x-y'|^N} \right| \, dx$$

$$= \int_{\{x:|x-y|>2|y-y'|\}} \left| \frac{\Omega(x-y) \cdot |x-y'|^N - \Omega(x-y') \cdot |x-y|^N}{|x-y|^N \cdot |x-y'|^N} \right| \, dx$$

$$\leq 2^N \int_{\{x:|x-y|>2|y-y'|\}} \left| \frac{\Omega(x-y)[|x-y'|^N - |x-y|^N]}{|x-y|^{2N}} \right|$$

$$+ \frac{|x-y|^N |\Omega(x-y) - \Omega(x-y')|}{|x-y|^{2N}} \, dx.$$

$$(6.2.11)$$

Here we have added and subtracted a convenient term in the numerator and we have used the fact that $|x-y|$ and $|x-y'|$ are comparable on the domain of integration. This last is true because

$$|x-y| \leq |x-y'| + |y-y'| \leq |x-y'| + \frac{|x-y|}{2};$$

hence

$$|x-y| \leq 2|x-y'|;$$

and a similar calculation shows that

$$|x-y'| \leq 2|x-y|.$$

The numerator of the first expression in (6.2.11) is easily estimated by $C|x - y|^{N-1} \cdot |y - y'|$. Recall that, in classical Calderón-Zygmund theory, we always assume that Ω is smooth on the unit sphere. Using this fact, we may estimate the numerator of the second term by a constant times

$$|x - y|^N \cdot |y - y'| \cdot |x - y|^{-1} \le |y - y'| \cdot |x - y|^{N-1}.$$

Altogether then, the last integral in displayed formula (6.2.11) is not greater than

$$C \cdot \int_{\{x:|x-y|>2|y-y'|\}} \frac{|y - y'|}{|x - y|^{N+1}} \, dx.$$

With a translation of coordinates, and setting $|y - y'| = \beta$, we may rewrite this last as

$$C \cdot \beta \cdot \int_{\{x:|x|>2\beta\}} \frac{1}{|x|^{N+1}} \, dx.$$

Polar coordinates enable us to write this as

$$C \cdot \beta \int_{\Sigma_{N-1}} \int_{2\beta}^{\infty} \frac{1}{r^{N+1}} \cdot r^{N-1} dr d\sigma.$$

Here Σ_{N-1} is the unit sphere in $\mathbb{R}^N$ and $d\sigma$ the standard rotationally invariant measure on that sphere. The factor r^{N-1} is the usual Jacobian that comes from the change from rectangular to spherical coordinates.

It is a trivial calculus exercise to see that this last displayed expression is majorized by a constant. Thus we see that the classical Calderón-Zygmund kernels satisfy the more subtle Hörmander cancellation condition.

6.3   A New Look at Hardy Spaces

Recall from Section 5.6 that an element of the Hardy space H^1 was defined to be an L^1 function f with the property that $R_j f \in L^1$, $j = 1, \ldots, N$—here $R_1, \ldots, R_N$ are the Riesz transforms. This definition, while natural from the partial differential equations or integral transform point of view, is rather cumbersome to treat in practice. For this reason, and for the more far-reaching purpose of freeing the notion of Hardy space from the rather rigid structure of Euclidean space, we formulated in Section 5.7 a new definition of the Hardy spaces in terms of atoms. Of course, this atomic formulation of the Hardy space theory is, in the context of $\mathbb{R}^N$, completely equivalent to the formulation in terms of Riesz transforms or the formulation in terms of maximal functions (Section 5.6).

On a space of homogeneous type, there is not sufficient structure to consider an analogue of the Riesz transforms. There is, in general, no notion of "smooth" function. Indeed, there is not even enough structure to define convolution; so it is not immediately clear how to consider a maximal function definition of Hardy spaces (although see work of Macias-Segovia [MAS], Uchiyama [UCH], and Han [HAN]). Thus, on a general space of homogeneous type, the one and only universal definition of the Hardy spaces is in terms of atoms.

Let X be a space of homogeneous type as usual. Let a be an L^1 function. We say that a is a 1-atom (or, as is sometimes seen in the literature, a $(1, \infty)$-atom) if

(6.3.1) $\operatorname{supp} a \subseteq B(x, r)$ for some $x \in X, r > 0$;

(6.3.2) $|a(t)| \leq \dfrac{1}{\mu(B(x, r))}$ for all $t \in B(x, r)$;

(6.3.3) $\displaystyle\int_{B(x,r)} a(t) \, d\mu(t) = 0 \,.$

Refer to Section 5.7 to see how the atomic theory gave us a third characterization (after the Riesz transforms and the maximal functions) of Hardy spaces on Euclidean space. Now that we are in a setting where we cannot make any sense of Riesz transforms nor of maximal functions defined by convolution, we use the third method to *define* Hardy space functions in circumstances where the usual artifacts of Euclidean structure are not present:

Definition 6.3.4 Let (X, ρ, μ) be a space of homogeneous type. Define a 1-atom as above (noticing that a space of homogeneous type has all the structure necessary to formulate the three axioms for an atom). Then any function of the form

$$f = \sum_j \beta_j a_j,$$

where $\sum_j |\beta_j| < \infty$ and each a_j is an atom,[3] is called a *real-variable* or *atomic H^1* function. We write $f \in H^1_{\mathrm{Re}}$.

Let us prove a sample theorem to illustrate the utility of atoms. This result is actually quite difficult to prove if one uses the classical definition of H^1. By using the atomic approach, we get a much more transparent proof and also one that is valid in the rather general setting of spaces of homogeneous type.

Theorem 6.3.5 (Fefferman, Stein) *Let k be a Calderón-Zygmund singular integral kernel. Then the operator T_k maps H^1_{Re} to L^1.*

Remark. In fact more is true, at least on $\mathbb{R}^N$. The real theorem is that T_k maps $H^1_{\mathrm{Re}}(\mathbb{R}^N)$ to $H^1_{\mathrm{Re}}(\mathbb{R}^N)$, but this assertion is rather more tedious to prove (see [FES], [KRA2]), even with the advantages of the atomic approach. Notice that the full theorem serves to vindicate the importance of the real-variable Hardy spaces: H^1 serves as a useful substitute for L^1 (in the context of integral operators), because singular integrals and fractional integrals are not bounded on L^1 but they are bounded on H^1.

Notice how the proof that follows uses both the L^2 boundedness hypothesis and the Hörmander condition in Theorem 6.2.10 quite explicitly. Let the constants M_1, M_2, M_3 be as in 6.2.10.

---

[3] The convergence here is obviously in L^1, although the analogous result for H^p, $p < 1$ involves a more subtle form of convergence in the so-called distribution topology.

Proof of the Theorem. Our job is to estimate

$$\int_X |T_k a| \, d\mu = \int_X \left| \int_X k(x, y) a(y) \, d\mu(y) \right| \, d\mu(x). \qquad (6.3.5.1)$$

We will derive an estimate for this iterated integral that is independent of the particular atom a. Since any H^1 function is a linear combination of atoms with summable coefficients, the result will follow.

Now let us rewrite (6.3.5.1) as

$$\int_{B(P, M_2 r)} \left| \int_X k(x, y) a(y) \, d\mu(y) \right| \, d\mu(x)$$

$$+ \int_{X \setminus B(P, M_2 r)} \left| \int_X k(x, y) a(y) \, d\mu(y) \right| \, d\mu(x) \equiv I + II,$$

where $B(P, r)$ contains the support of the atom a as in the axioms for an atom. We estimate term I by noticing that a is certainly an L^2 function and T_k is bounded on L^2. Thus

$$|I| \le \|\chi_{B(P, M_2 r)}\|_{L^2} \cdot \|T_k a\|_{L^2}$$

$$\le [\mu(B(P, M_2 r))]^{1/2} \cdot M_1 \|a\|_{L^2}$$

$$\le M_1 \cdot [\mu(B(P, M_2 r))]^{1/2} \cdot [\mu(B(P, r))]^{-1/2}$$

$$\le C'.$$

That completes our analysis of term I.

For term II, we invoke the fact that a has mean zero to rewrite the expression as

$$\int_{X \setminus B(P, M_2 r)} \left| \int_X [k(x, y) - k(x, P)] a(y) \, d\mu(y) \right| \, d\mu(x).$$

Now we pull the supremum norm of a out in front of the integral to estimate this last expression as

$$\frac{1}{\mu(B(P, M_2 r))} \int_{X \setminus B(P, M_2 r)} \int_{B(P, r)} |k(x, y) - k(x, P)| \, d\mu(y) \, d\mu(x).$$

Now the Hörmander condition gives the desired estimate, independent of the choice of a. $\qquad\qquad\qquad\qquad\qquad\qquad\qquad\qquad\qquad\qquad$ $\square$

In Section 5.7, we briefly discussed the theory of atomic H^p for $p < 1$. As p becomes smaller, higher-order moment conditions are involved and the theory becomes rather complex. Maximal functions and atoms serve to bypass much of that complexity, and are valuable tools.

It is only a recent development that techniques have been devised for defining Hardy spaces for small p on a space of homogeneous type (see [KRL]). We shall say nothing about these new ideas here.

6.4   The $T(1)$ Theorem

Prior to the $T(1)$ theorem, there were precious few methods for determining whether a non-convolution operator is bounded on L^2. The celebrated $T(1)$ theorem addresses this issue head on, and provides a powerful new tool. Modern variants of the theorem apply to the other standard function spaces of harmonic analysis—the L^p spaces, Lipschitz spaces, Hardy spaces, Sobolev spaces, Orlicz spaces, and so forth. In this brief exposition we shall concentrate on the original L^2 theory.

Let us motivate the David-Journé theorem, commonly known as the $T(1)$ theorem, by examining the classical Calderón-Zygmund singular integral operators that are already familiar to us. We are endeavoring to find conditions on a more general kernel $k(x, y)$ which will make the operator

$$T_k : f \longmapsto \int k(x, y) f(y) \, dy \qquad\qquad (6.4.1)$$

bounded on L^2. Our aim is to first examine the situation for a convolution kernel $K(x)$, and the associated operator

$$f \longmapsto \int K(x - y) f(y) \, dy, \qquad\qquad (6.4.2)$$

which we already understand.

The most standard method for seeing that the operator (6.4.2) is bounded on L^2 is to show that $\widehat{K}$ is a bounded function. A moment's discussion is merited. First, the classical Calderón-Zygmund kernel K is *not* locally integrable. However, the mean-value-zero property of such kernels guarantees that integration against K makes sense on C_c^∞ functions (see the argument in Section 5.2). In other words, integration against K (suitably interpreted) is a distribution. [The reader unfamiliar with distribution theory should not get nervous at this point. A willing suspension of disbelief will see such a reader through, and something should be gained along the way.] In fact it is a Schwartz distribution, so we may consider its Fourier transform (see Appendix II).

Now the distribution given by integration against K is homogeneous of degree $-N$ (because the function K is homogeneous of degree $-N$ and has mean-value zero), hence the Fourier transform of this distribution will be homogeneous of degree 0 (see Section 2.2). In fact, with some extra effort, it can be seen to be a bounded function. That is what we wished to see.

The arguments just given cannot apply to an operator given by integration against a general kernel $k(x, y)$ because such an operator is not translation-invariant. There are alternative methods for seeing that an operator is bounded on L^2, and these sometimes apply quite nicely in situations where there is no usable Fourier transform, or where the operator in question is not translation-invariant. We now briefly describe one of the most important of these.

Let $T_1, T_2, \ldots, T_M$ be bounded operators from a Hilbert space H to itself, and assume that each T_j has operator norm 1. What can we say about the operator norm of $T = \sum_{j=1}^{M} T_j$? In general, the answer is that T has operator norm not exceeding M. In fact, if each T_j were the identity operator, then the operator norm of T would be M and that would be the end of that.

It is a deep insight of Mischa Cotlar [COT] that if the operators T_j *act on different parts of the Hilbert space H*, then when the operators are summed, they in fact do not superimpose. Here is a simple example. Suppose that H is the Hilbert space $L^2(\mathbb{T})$. Let

$$T_j f = f * e_j = \widehat{f}(j) \cdot e_j,$$

where $e_j(x) = e^{ijx}$. Then of course each operator T_j has norm precisely 1. But Bessel's inequality (Section 0.4) tells us that any operator $T = \sum_{j=1}^{M} T_j$ also has norm 1. And we see that, in a palpable sense, each T_j sends the 1-dimensional space in H_j that is spanned by e_j to itself (indeed it is the projection onto that space). Cotlar's insight was to determine how to abstract this setup to a general theorem about summing Hilbert space operators.[4]

Cotlar's theorem [COT] is elegant and beautiful, and has a wonderfully intricate proof, but its hypotheses are so restrictive as to render it difficult to use in applications. In particular, if we want to study an operator T by decomposing it into operators T_j, with $T = \sum_j T_j$, then it is required, in the hypotheses of Cotlar's theorem, that each T_j be self-adjoint and that the operators T_j commute. These hypotheses indeed hold in the example of the last paragraph, but it is difficult to force this circumstance in a typical application.

Some time after [COT] appeared, Cotlar himself (unpublished) and Knapp-Stein [KNS2] independently discovered a very natural generalization of the original Cotlar theorem. [The rather interesting history of this theorem, and its various formulations, can be found in [COT], [KNS1], and [STE2].] For cultural purposes we state it here, but we do not prove it. What is now called the Cotlar-Knapp-Stein theorem has proved to be a powerful and versatile tool in harmonic analysis. It also served as a role model (and as a key ingredient in the proof) for

---

[4]Another, slightly more abstract, way to look at these matters is as follows. Let the Hilbert space H be the orthogonal direct sum of Hilbert spaces H_j. For each j, let T_j be an operator that maps H_j to itself and maps $H_j^{\perp}$ to 0. Suppose that, for each j, $\|T_j\|_{\text{op}} \leq C$. Now if $x \in H$ we write $x = \sum x_j$ with $x_j \in H_j$ and we calculate that

$$\left(\sum T_j\right) x = \sum T_j x_j$$

so that

$$\left\|\sum T_j x\right\|^2 = \sum \|T_j x_j\|^2 \leq C^2 \sum \|x_j\|^2 = C^2 \|x\|^2.$$

the sort of abstract result about L^2 boundedness of integral operators that David-Journé ultimately proved.

Theorem 6.4.3 (Cotlar-Knapp-Stein) *Let H be a Hilbert space and let $T_j : H \to H$, $j \geq 1$, be bounded operators, with $\|T_j^* T_k\|_{\mathrm{op}} \leq \lambda(j - k)$ and $\|T_j T_k^*\|_{\mathrm{op}} \leq \lambda(k - j)$ for some positive function λ and all j, k. Assume that there is a constant $C > 0$ such that*

$$\sum_j \lambda(j)^{1/2} = C < \infty.$$

[Notice that when $j = k$ we get a hypothesized uniform bound of $\sqrt{\lambda(0)}$ on the norms of the $T_j s$.] Then, for any positive integer M,

$$\left\| \sum_{j=1}^M T_j \right\|_{\mathrm{op}} \leq C.$$

Notice that the bound given for $\| \sum_j T_j \|_{\mathrm{op}}$ is independent of M. It is possible to use this theorem to prove that the Hilbert transform is bounded on L^2. Indeed, one takes the classical kernel $K(x) = 1/x$ and breaks it up by defining

$$K_j(x) = \begin{cases} K(x) & \text{if} \quad 2^j < |x| \leq 2^{j+1} \\ 0 & \text{if} \quad 2^{j+1} < |x| \text{ or } |x| \leq 2^j, \end{cases}$$

$-\infty < j < \infty$. If T_j is the operator given by convolution against K_j, then it can be shown that $\|T_j^* T_k\|_{\mathrm{op}} \leq C \cdot 2^{-(j-k)}$ and $\|T_j T_k^*\|_{\mathrm{op}} \leq C \cdot 2^{-(j-k)}$. Thus the Cotlar-Knapp-Stein theorem applies and any finite sum

$$\sum_{j=-M}^N T_j$$

has operator norm that can be bounded independent of M and N. Then a limiting argument gives the full result that the original operator—the Hilbert transform—is bounded on L^2.

It turns out, as we shall see below, that the Cotlar-Knapp-Stein theorem is a logical stepping stone to the $T(1)$ theorem.

Now let us return to the $T(1)$ theorem. Although there are versions of the $T(1)$ theorem for a space of homogeneous type, it is simpler in a first pass for us to restrict attention to Euclidean space. So fix a Euclidean space $\mathbb{R}^N$. Let $\phi \in C_c^K(\mathbb{R}^N)$, some $K > 0$ (where K is usually specified in context). For $\epsilon > 0$ and $x \in \mathbb{R}^N$ we set

$$\phi^{x,\epsilon}(y) = \phi((y - x)/\epsilon).$$

Definition 6.4.4 We say that an integral operator T_k is *weakly bounded* if, for each bounded subset S of $C_c^K(\mathbb{R}^N)$, there is a constant $C = C_S$ such that

$$\left| \int (T_k \phi^{x,\epsilon})(s) \psi^{x,\epsilon}(s)\, ds \right| \leq C \cdot \epsilon^N$$

for all $\epsilon > 0$, all $x \in \mathbb{R}^N$, and all $\phi, \psi \in S$. [Here K is chosen to be sufficiently large, depending on the dimension N of $\mathbb{R}^N$.]

Observe that

$$\|\phi^{x,\epsilon}\|_{L^2} \cdot \|\psi^{x,\epsilon}\|_{L^2} \leq C \|\phi\|_{L^2} \|\psi\|_{L^2} \cdot \epsilon^N$$

by a simple change of variables. This estimate suggests that L^2 boundedness is a stronger condition than weak boundedness. What we seek now is a converse implication.

An implicit aspect of the theory is that we restrict attention to kernels that have a standard form. That form is related to the decay of the kernel and its derivatives at infinity. Here is one formulation (following Christ [CHR]) of these conditions:

Definition 6.4.5 We say that a kernel $k(x, y)$ on $\mathbb{R}^N \times \mathbb{R}^N$ is a *standard kernel* if, for some $\delta > 0$ and $0 < C < \infty$ and for all $x, y, z \in \mathbb{R}^N$ such that $|x - z| < |x - y|/2$, we have

(6.4.5.1) $|k(x, y)| \le C|x - y|^{-N}$;

(6.4.5.2) $|k(x, y) - k(z, y)| \le C \left(\dfrac{|x - z|}{|x - y|} \right)^{\delta} \cdot |x - y|^{-N}$;

(6.4.5.3) $|k(y, x) - k(y, z)| \le C \left(\dfrac{|x - z|}{|x - y|} \right)^{\delta} \cdot |x - y|^{-N}$.

A moment's thought reveals that these conditions for a standard kernel are at least partly inspired by the "Hörmander condition" on a singular integral kernel in Theorem 6.2.10.

Here is a classical statement of the celebrated $T(1)$ theorem. In its statement, we let **1** denote the constant function that is identically equal to 1.

Theorem 6.4.6 (David-Journé) *Let $k(x, y)$ be a standard kernel. Assume that the operator*

$$T : f \longmapsto \int f(y)k(x, y) \, dy$$

satisfies

(6.4.6.1) *T is weakly bounded;*

(6.4.6.2) *$T(1) \in BMO$;*

(6.4.6.3) *$T^t(1) \in BMO$; [Here T^t is the usual transpose of T—see [CHR].]*

Then T is bounded on L^2. The converse is true as well.

We shall not give a detailed proof of the $T(1)$ theorem, but shall instead content ourselves with some remarks. First, what is the sense of $T(1)$? The simplest way to think about such an expression is to fix a smooth cutoff function ϕ that is identically equal to 1 near the origin. Let $\phi_\delta(x) = \phi(\delta x)$. Then we may consider, for $\delta > 0$ fixed and for $\epsilon > 0$, the expression

$$I_\epsilon(\phi_\delta \mathbf{1})(x) \equiv \int_{|x-y|>\epsilon} k(x, y)[\phi_\delta(y) \cdot \mathbf{1}(y)] \, dy.$$

Since k is a standard kernel, the integral converges for each fixed ϵ, δ. To make sense of $T(1)$, we first require that $\lim_{\epsilon \to 0^+} I_\epsilon(\phi_\delta 1)$ exist as a function. Then we let $\delta \to 0^+$. The resulting function should be in BMO. There are other, more canonical ways to think about this matter, and we refer the reader to [CHR] for the details.

It is fairly simple to verify that a classical Calderón-Zygmund operator maps 1 to the BMO function that is identically 0. The kernel of such an operator also satisfies the estimates for a standard kernel, and is weakly bounded virtually by inspection. So the $T(1)$ theorem subsumes the classical Calderón-Zygmund theorem.

The proof of the $T(1)$ theorem entails subtracting from T two so-called paraproduct operators that are modeled on $B \equiv T(1)$ and $b' \equiv T'(1)$. [Paraproducts, a construct of J. Bony, are a standard part of pseudodifferential operator theory—see [STR2].] What is left behind is an operator to which the Cotlar-Knapp-Stein theorem may be applied. Details may be found in [CHR].

We conclude by addressing the following question: "What is so special about the function 1? Are there not other functions that we could test?" The answer is "yes," and the functions that most naturally may be substituted for 1 are the para-accretive functions. A function b is called *para-accretive* if it is bounded and is also essentially bounded away from zero, in a measure-theoretic sense. The theorem obtained when the "testing function" 1 is replaced by a para-accretive function b is the so-called $T(b)$ theorem of David-Journé-Semmes. We again refer the reader to [CHR] for details and background.

The $T(1)$ and $T(b)$ theorems not only put the Calderón-Zygmund theory on a more natural footing, but they provide a large family of operators that are now easily checked to be bounded on L^2 and which cannot be handled by classical methods. There are profound applications to Calderón commutator theory, to singular integrals on Lipschitz curves, to operators that arise in the harmonic analysis of several complex variables, and to partial differential equations. We shall say no more about these matters here.

CHAPTER 7

Wavelets

7.1 Localization in the Time and Space Variables

The premise of the new versions of Fourier analysis that are being developed today is that sines and cosines are not an optimal model for some of the phenomena that we want to study. As an example, suppose that we are developing software to detect certain erratic heartbeats by analysis of an electrocardiogram. [Note that the discussion that we present here is philosophically correct but is oversimplified to facilitate the exposition.] The scheme is to have the software break down the patient's electrocardiogram into component waves. If a wave that is known to be a telltale signal of heart disease is detected, then the software notifies the user.

A good plan, and there is indeed software of this nature in use across America. But let us imagine that a typical electrocardiogram looks like that shown in Figure 1. Imagine further that the aberrant heartbeat that we wish to detect is the one in Figure 2.

What we want the software to do is to break up the wave in Figure 1 into fundamental components, and then to see whether one of those components is the wave in Figure 2. Of what utility is Fourier theory in such an analysis? Fourier theory would allow us to break the wave in Figure 1 into sines and cosines, then break the wave in Figure 2

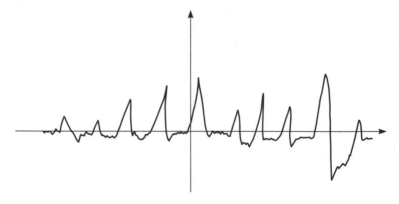

Figure 1. An electrocardiogram.

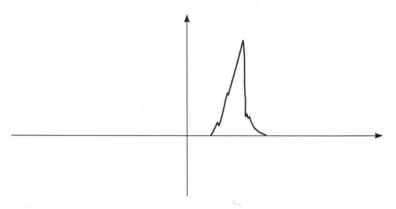

Figure 2. An aberrant heartbeat.

into sines and cosines, and then attempt to match up coefficients. Such a scheme may be dreadfully inefficient, because sines and cosines *have nothing to do* with the waves we are endeavoring to analyze.

The Fourier analysis of sines and cosines arose historically because sines and cosines are eigenfunctions for the wave equation (see Chapters 0 and 1). Their place in mathematics became even more firmly secured because they are orthonormal in L^2. They also commute with translations in natural and useful ways. The standard trigonometric relations between the sine and cosine functions give rise to elegant and useful formulas—such as the formulas for the Dirichlet kernel and the Fejér kernel and the Poisson kernel. Sines and cosines have played an inevitable and fundamental historical role in the development of harmonic analysis.

In the same vein, translation-invariant operators have played an important role in our understanding of how to analyze partial differential equations (see [KRA3]), and as a step toward the development of the more natural theory of pseudodifferential operators. Today we find ourselves studying translation *non*invariant operators—such as those that arise in the analysis on the boundary of a (smoothly bounded) domain in $\mathbb{R}^N$ (see Figure 3). The $T(1)$ theorem of David-Journé (Section 6.4) gives the most natural and comprehensive method of analyzing integral operators and their boundedness on a great variety of spaces.

The next, and current, step in the development of Fourier analysis is to replace the classical sine and cosine building blocks with more flexible units—indeed, with units that can be tailored to the situation at hand. Such units should, ideally, be localizable; in this way they can more readily be tailored to any particular application. This, roughly speaking, is what wavelet theory is all about.

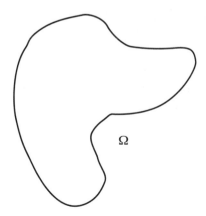

Figure 3.   A smoothly bounded domain.

In a book of this nature, we clearly cannot develop the full assemblage of tools that are a part of modern wavelet theory. [See [HERG], [MEY1], [MEY2], and [DAU] for more extensive treatments of this beautiful and dynamic subject. The papers [STR3] and [WAL] provide nice introductions as well.] What we can do is to give the reader a taste. Specifically, we shall develop a Multi-Resolution Analysis, or MRA; this study will show how Fourier analysis may be carried out with localization in either the space variable or the Fourier transform (frequency) variable. In short, the reader will see how either variable may be localized. Contrast this notion with the classical construction, in which the units are sines and cosines—clearly functions that *do not* have compact support. The exposition here derives from that in [HERG], [STR3], and [WAL]. We also thank G. B. Folland and J. Walker for considerable guidance in preparing this chapter.

7.2   Building a Custom Fourier Analysis

Typical applications of classical Fourier analysis are to

- **Frequency Modulation:** Alternating current, radio transmission;
- **Mathematics:** Ordinary and partial differential equations, analysis of linear and nonlinear operators;
- **Medicine:** Electrocardiography, magnetic resonance imaging, biological neural systems;
- **Optics and Fiber-Optic Communications:** Lens design, crystallography, image processing;
- **Radio, Television, Music Recording:** Signal compression, signal reproduction, filtering;
- **Spectral Analysis:** Identification of compounds in geology, chemistry, biochemistry, mass spectroscopy;
- **Telecommunications:** Transmission and compression of signals, filtering of signals, frequency encoding.

In fact, the applications of Fourier analysis are so pervasive that they are part of the very fabric of modern technological life.

The applications that are being developed for wavelet analysis are very similar to those just listed. But the wavelet algorithms give rise to faster and more accurate image compression, faster and more accurate signal compression, and better denoising techniques that preserve the original signal more completely. The applications in mathematics lead, in many situations, to better and more rapid convergence results.

What is lacking in classical Fourier analysis can be readily seen by examining the Dirac delta mass. Because, if the unit ball of L^1— thought of as a subspace of the dual space of $C(\mathbb{T})$—had any extremal functions (it does not), they would be objects of this sort: the weak-$*$ limit of functions of the form $N^{-1}\chi_{[-1/2N,1/2N]}$ as $N \to +\infty$. That weak-$*$ limit is the Dirac mass. We know the Dirac mass as the functional that assigns to each continuous function with compact support its value at 0:

$$\delta : C_c(\mathbb{R}^N) \ni \phi \longmapsto \phi(0).$$

It is most convenient to think of this functional as a measure:

$$\int \phi(x)\,d\delta(x) = \phi(0).$$

Now suppose that we want to understand δ by examining its Fourier transform. For simplicity, restrict attention to $\mathbb{R}^1$:

$$\widehat{\delta}(\xi) = \int_{\mathbb{R}} e^{i\xi \cdot t}\,d\delta(t) \equiv 1.$$

In other words, the Fourier transform of δ is the constant, identically 1, function. To recover δ from its Fourier transform, we would have to make sense of the inverse Fourier integral

$$\int 1 \cdot e^{-i\xi \cdot t}\,dt.$$

Doing so requires a careful examination of the Gauss-Weierstrass summation method, and certainly strains the intuition: why should we have to "sum" exponentials, each of which is supported on the entire line and none of which is in any L^p class for $1 \leq p < \infty$, in order to reconstruct δ—which is supported just at the origin?

The point comes through perhaps even more strikingly by way of Fourier series. Consider the Dirac mass δ supported at the origin in the circle group $\mathbb{T}$. Then the Fourier-Stieltjes coefficients of δ are

$$\widehat{\delta}(j) \equiv \frac{1}{2\pi} \int_{-\pi}^{\pi} e^{-ijt} \, d\delta(t) = 1.$$

Thus recovering δ from its Fourier series amounts to finding a way to sum the formal series

$$\sum_{j=-\infty}^{\infty} 1 \cdot e^{ijt}$$

in order to obtain the Dirac mass. Since each exponential is supported on the entire circle group, the imagination is defied to understand how these exponentials could sum to a point mass. [To be fair, the physicists

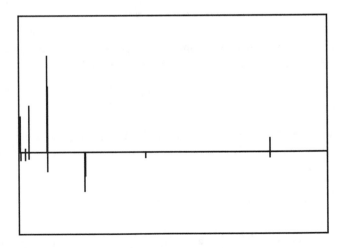

Figure 1.   Noise spikes as point masses.

have no trouble seeing this point: at the origin the terms all add up, and away from zero they all cancel out.]

The study of the point mass is not merely an affectation. In a radio signal, noise (in the form of spikes) is frequently a sum of point masses (Figure 1). On a phonograph record, the pops and clicks that come from imperfections in the surface of the record exhibit themselves (on an oscilloscope, for instance) as spikes, or point masses.

For the sake of contrast, in the next sections we shall generate an *ad hoc* family of wavelet-like basis elements for L^2 and show how these may be used much more efficiently to decompose the Dirac mass into basis elements.

7.3   The Haar Basis

In this section we shall describe the Haar wavelet basis. While the basis elements are not smooth functions (as wavelet basis elements usually are), they will exhibit the other important features of a Multi-Resolution Analysis (MRA). In fact, we shall follow the axiomatic treatment as developed by Mallat and exposited in [WAL] in order to isolate the essential properties of an MRA.

We shall produce a dyadic version of the wavelet theory. Certainly other theories, based on other dilation paradigms, may be produced. But the dyadic theory is the most standard, and quickly gives the flavor of the construction. In this discussion we shall use, as we did in Chapter 2, the notation α_δ to denote the dilate of a function: $\alpha_\delta f(x) \equiv f(\delta x)$. And we shall use the notation τ_a to denote the translate of a function: $\tau_a f(x) \equiv f(x - a)$.

We work on the real line $\mathbb{R}$. Our universe of functions will be $L^2(\mathbb{R})$. Define

$$\phi = \chi_{[0,1)}$$

and

$$\psi(x) \equiv \phi(2x) - \phi(2x - 1) = \chi_{[0,1/2)}(x) - \chi_{[1/2,1)}(x).$$

These functions are exhibited in Figure 1.

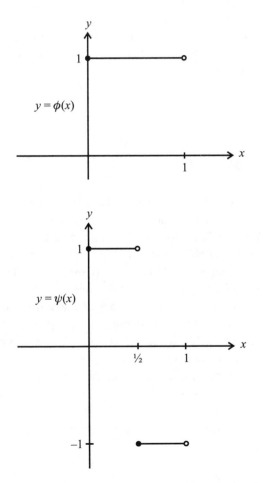

Figure 1. The scaling function ϕ and the wavelet function ψ.

The function ϕ will be called a *scaling function* and the function ψ will be called the associated *wavelet*. The basic idea is this: translates of ϕ will generate a space V_0 that can be used to analyze a function f on a large scale—more precisely, on the scale of size 1 (because 1 is the length of the support of ϕ). But the elements of the space V_0 cannot be used to detect information that is at a scale *smaller* than 1. So we will scale the elements of V_0 down by a factor of 2^j, each $j = 1, 2, \ldots$, to obtain a space that can be used for analysis at the scale 2^{-j} (and we will also scale V_0 *up* to obtain elements that are useful at an arbitrarily large scale). Let us complete this program now for the specific ϕ that we have defined above, and then present some axioms that will describe how this process can be performed in a fairly general setting.

Now we use ϕ to generate a scale of function spaces $\{V_j\}_{j \in \mathbb{Z}}$. We set

$$V_0 = \left\{ \sum_{k \in \mathbb{Z}} a_k [\tau_k \phi] : \sum |a_k|^2 < \infty \right\},$$

for the particular function ϕ that was specified above. Of course each element of V_0 so specified lies in L^2 (because the functions $\tau_k \phi$ have disjoint supports). But it would be wrong to think that V_0 is all of L^2, for an element of V_0 is constant on each interval $[k, k+1)$ and has possible jump discontinuities only at the integers. The functions $\{\tau_k \phi\}_{k \in \mathbb{Z}}$ form an orthonormal basis (with respect to the L^2 inner product) for V_0.

Now let us say that a function g is in V_1 if and only if $\alpha_{1/2}(g)$ lies in V_0. Thus $g \in V_1$ means that g is constant on the intervals determined by the lattice $(1/2)\mathbb{Z} \equiv \{n/2 : n \in \mathbb{Z}\}$ and has possible jump discontinuities only at the elements of $(1/2)\mathbb{Z}$. It is easy to see that the functions $\{\sqrt{2}\alpha_2 \tau_k \phi\}$ form an orthonormal basis for V_1.

Observe that $V_0 \subseteq V_1$ since every jump point for elements of V_0 is also a jump point for elements of V_1 (but not conversely). More explicitly, we may write

$$\tau_k \phi = \alpha_2 \tau_{2k} \phi + \alpha_2 \tau_{2k+1} \phi,$$

thus expressing an element of V_0 as a linear combination of elements of V_1.

Now that we have the idea down, we may iterate it to define the spaces V_j for any $j \in \mathbb{Z}$. Namely, for $j \in \mathbb{Z}$, V_j will be generated by the functions $\alpha_{2^j} \tau_m \phi$, all $m \in \mathbb{Z}$. In fact we may see explicitly that an element of V_j will be a function of the form

$$f = \sum_{\ell \in \mathbb{Z}} a_\ell \chi_{[\ell/2^j, [\ell+1]/2^j)}$$

where $\sum |a_\ell|^2 < \infty$. Thus an orthonormal basis for V_j is given by $\{2^{j/2} \alpha_{2^j} \tau_m \phi\}_{m \in \mathbb{Z}}$.

Now the spaces V_j have no common intersection except the zero function. This is so because, since a function $f \in \cap_{j \in \mathbb{Z}} V_j$ would be constant on arbitrarily large intervals (of length 2^{-j} for j negative), then it can only be in L^2 if it is zero. Also, $\cup_{j \in \mathbb{Z}} V_j$ is dense in L^2 because any L^2 function can be approximated by a simple function (i.e., a finite linear combination of characteristic functions), and any characteristic function can be approximated by a sum of characteristic functions of dyadic intervals.

We therefore might suspect that if we combine all the orthonormal bases for all the V_j, $j \in \mathbb{Z}$, then this would give an orthonormal basis for L^2. That supposition is, however, incorrect. For the basis elements $\phi \in V_0$ and $\alpha_{2^j} \tau_0 \phi \in V_j$ are not orthogonal. This is where the function ψ comes in.

Since $V_0 \subseteq V_1$, we may proceed by trying to complete the orthonormal basis $\{\tau_k \phi\}$ of V_0 to an orthonormal basis for V_1. Put in other words, we write $V_1 \equiv V_0 \oplus W_0$, and we endeavor to write a basis for W_0. Let $\psi = \alpha_2 \phi - \alpha_2 \tau_1 \phi$ be as above, and consider the set of functions $\{\tau_m \psi\}$. Then this is an orthonormal set. Let us see that it spans W_0.

Let h be an arbitrary element of W_0. So certainly $h \in V_1$. It follows that

$$h = \sum_j b_j \alpha_2 \tau_j \phi$$

for some constants $\{b_j\}$ that are square-summable. Of course h is constant on the interval $[0, 1/2)$ and also constant on the interval $[1/2, 1)$. We note that

$$\phi(t) = \frac{1}{2} [\phi(t) + \psi(t)] \qquad \text{on } [0, 1/2)$$

and

$$\phi(t) = \frac{1}{2} [\phi(t) - \psi(t)] \qquad \text{on } [1/2, 1).$$

It follows that

$$h(t) = \left(\frac{b_0 + b_1}{2} \right) \phi(t) + \left(\frac{b_0 - b_1}{2} \right) \psi(t)$$

on $[0, 1)$. A similar decomposition obtains on every interval $[j, j + 1)$. As a result,

$$h = \sum_{j \in \mathbb{Z}} c_j \tau_j \phi + \sum_{j \in \mathbb{Z}} d_j \tau_j \psi,$$

where

$$c_j = \frac{b_j + b_{j+1}}{2} \qquad \text{and} \qquad d_j = \frac{b_j - b_{j+1}}{2}.$$

Note that $h \in W_0$ implies that $h \in V_0^\perp$. Also, every $\tau_j \phi$ is orthogonal to every $\tau_k \psi$. Consequently every coefficient $c_j = 0$. Thus we have proved that h is in the closed span of the terms $\tau_j \psi$. In other words, the functions $\{\tau_j \psi\}_{j \in \mathbb{Z}}$ span W_0.

Thus we have $V_1 = V_0 \oplus W_0$, and we have an explicit orthonormal basis for W_0. Of course we may scale this construction up and down to obtain

$$V_{j+1} = V_j \oplus W_j \qquad (7.3.1)_j$$

for every j. And we have the explicit orthonormal basis $\{2^{j/2}\alpha_{2^j}\tau_m\psi\}_{m\in\mathbb{Z}}$ for each W_j.

We may iterate the equation $(7.3.1)_j$ to obtain

$$V_{j+1} = V_j \oplus W_j = V_{j-1} \oplus W_{j-1} \oplus W_j$$
$$= \cdots = V_0 \oplus W_0 \oplus W_1 \oplus \cdots \oplus W_{j-1} \oplus W_j.$$

Letting $j \to +\infty$ yields

$$L^2 = V_0 \oplus \bigoplus_{j=0}^{\infty} W_j. \tag{7.3.2}$$

But a similar decomposition may be performed on V_0, with W_j in descending order:

$$V_0 = V_{-1} \oplus W_{-1} = \cdots = V_{-\ell} \oplus W_{-\ell} \oplus \cdots \oplus W_{-1}.$$

Letting $\ell \to +\infty$, and substituting the result into (7.3.2), now yields that

$$L^2 = \bigoplus_{j\in\mathbb{Z}} W_j.$$

Thus we have decomposed $L^2(\mathbb{R})$ as an orthonormal sum of Haar wavelet subspaces. We formulate one of our main conclusions as a theorem:

Theorem 7.3.3 *The collection*

$$\mathcal{H} \equiv \left\{\alpha_{2^j}\tau_m\psi : m, j \in \mathbb{Z}\right\}$$

is an orthonormal basis for L^2, and will be called a wavelet basis for L^2.

Now it is time to axiomatize the construction that we have just performed in a special instance.

Axioms for a Multi-Resolution Analysis (MRA)

A collection of subspaces $\{V_j\}_{j \in \mathbb{Z}}$ of $L^2(\mathbb{R})$ is called a *Multi-Resolution Analysis* or MRA if

MRA_1	(Scaling)	For each j, the function $f \in V_j$ if and only if $\alpha_2 f \in V_{j+1}$;
MRA_2	(Inclusion)	For each j, $V_j \subseteq V_{j+1}$;
MRA_3	(Density)	The union of the V_j's is dense in L^2:

$$\text{closure} \left\{ \bigcup_{j \in \mathbb{Z}} V_j \right\} = L^2(\mathbb{R});$$

MRA_4	(Maximality)	The spaces V_j have no nontrivial common intersection:

$$\bigcap_{j \in \mathbb{Z}} V_j = \{0\};$$

MRA_5	(Basis)	There is a function ϕ such that $\{\tau_j \phi\}_{j \in \mathbb{Z}}$ is an orthonormal basis for V_0.

We invite the reader to review our discussion of $\phi = \chi_{[0,1)}$ and its dilates and confirm that the spaces V_j that we constructed do indeed form an MRA. Notice in particular that, once the space V_0 has been defined, then the other V_j are completely and uniquely determined by the MRA axioms.

7.4   Some Illustrative Examples

In this section we give two computational examples that provide concrete illustrations of how the Haar wavelet expansion is better behaved—especially with respect to detecting *local* data—than the Fourier series expansion.

Example 7.4.1 Our first example is quick and dirty. In particular, we cheat a bit on the topology to make a simple and dramatic point. It is this: if we endeavor to approximate the Dirac delta mass δ with a Fourier series, then the partial sums will always have a *slowly decaying* tail that extends far beyond the highly localized support of δ. By contrast, the partial sums of the Haar series for δ localize rather nicely. We will see that the Haar series has a tail too, but it is small.

Let us first examine the expansion of the Dirac mass in terms of the Haar basis. Properly speaking, what we have just proposed is not feasible because the Dirac mass does not lie in L^2. Instead let us consider, for $N \in \mathbb{N}$, functions

$$f_N = 2^N \chi_{[0,1/2^N)}.$$

The functions f_N each have mass 1, and the sequence $\{f_N\, dx\}$ converges, in the weak-$*$ sense of measures (i.e., the weak-$*$ topology—see Appendix VI), to the Dirac mass δ.

First, we invite the reader to calculate the ordinary Fourier series, or Fourier transform, of f_N (see also the calculations at the end of this example). Although (by the Riemann-Lebesgue lemma) the coefficients die out, the fact remains that any finite part of the Fourier transform, or any partial sum of the Fourier series, gives a rather poor approximation to f_N. After all, any partial sum of the Fourier series is a trigonometric polynomial, and any trigonometric polynomial has support on the *entire interval* $[0, 2\pi)$. In conclusion, whatever the merits of the approximation to f_N by the Fourier series partial sums, they are offset by the unwanted portion of the partial sum that exists *off the support of* f_N. [For instance, if we were endeavoring to construct a filter to remove pops and clicks from a musical recording, then the pop or click (which is mathematically modeled by a Dirac mass) would be replaced by the tail of a trigonometric polynomial—which amounts to undesired low level noise, as in Figure 1 below.]

Now let us do some calculations with the Haar basis. Fix an integer $N > 0$. If $j \geq N$, then any basis element for W_j will integrate to 0 on the support of f_N—just because the basis element will be 1 half the time and -1 half the time on each dyadic interval of length 2^{-j}. If

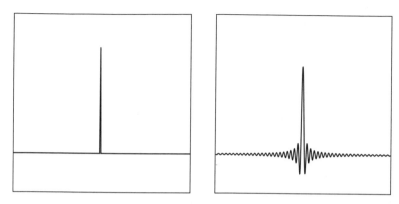

Figure 1. Undesired low level noise.

instead $j < N$, then the single basis element μ_j from W_j that has support intersecting the support of f_N is in fact constantly equal to $2^{j/2}$ on the support of f_N. Therefore the coefficient b_j of μ_j in the expansion of f_N is

$$b_j = \int f_N(x)\mu_j(x)\,dx = 2^N \int_0^{2^{-N}} 2^{j/2}\,dx = 2^{j/2}.$$

Thus the expansion for f_N is, for $0 \le x < 2^{-N}$,

$$\sum_{j=-\infty}^{N-1} 2^{j/2}\mu_j(x) = \sum_{j=-\infty}^{0} 2^{j/2} \cdot 2^{j/2} + \sum_{j=1}^{N-1} 2^{j/2} \cdot 2^{j/2}$$

$$= 2 + (2^N - 2)$$

$$= 2^N$$

$$= f_N(x).$$

Notice here that the contribution of terms of negative index in the series—which corresponds to "coarse scale" behavior that is of little interest—is constantly equal to 2 (regardless of the value of N) and is relatively trivial (i.e., small) compared to the interesting part of the series (of size $2^N - 2$) that comes from the terms of positive index.

If instead $2^{-N} \leq x < 2^{-N+1}$, then $\mu_{N-1}(x) = -2^{(N-1)/2}$ and $b_{N-1}\mu_{N-1}(x) = -2^{N-1}$; also

$$\sum_{j=-\infty}^{N-2} b_j \mu_j(x) = \sum_{j=-\infty}^{N-2} 2^j = 2^{N-1}.$$

Of course, $b_j = 0$ for $j \geq N$. In summary, for such x,

$$\sum_{j=-\infty}^{\infty} b_j \mu_j(x) = 0 = f_N(x).$$

A similar argument shows that if $2^{-\ell} \leq x < 2^{-\ell+1}$ for $-\infty < \ell \leq N$, then $\sum b_j \mu_j(x) = 0 = f_N(x)$. And the same result holds if $x < 0$.

Thus we see that the Haar basis expansion for f_N converges pointwise to f_N. More is true: the partial sums of the series give a rather nice approximation to the function f_N. Notice, for instance, that the partial sum $S_{N-1} = \sum_{j=-N+1}^{N-1} b_j \mu_j$ has the following properties:

(7.4.1.1) $S_{N-1}(x) = f_N(x) - 2^{-N+1}$ for $0 \leq x < 2^{-N}$;

(7.4.1.2) $S_{N-1}(x) = 0$ for $-2^{-N} < x < 0$;

(7.4.1.3) $S_{N-1}(x) = 0$ for $|x| > 2^N$;

(7.4.1.4) $|S_{N-1}(x)| \leq 2^{-N+1}$ for $2^{-N} \leq |x| \leq 2^N$.

It is worth noting that the partial sums of the Haar series for the Dirac mass δ,

$$H_N(x) = \sum_{|j| \leq N} 2^{j/2} \mu_j(x),$$

form (almost) a standard family of summability kernels as discussed in Chapter 1 (the missing feature is that each kernel integrates to 0 rather than 1); but the partial sums of the *Fourier* series for the Dirac mass δ,

$$D_N(x) = \sum_{|j| \leq N} 1 \cdot e^{ijx},$$

do *not* (as we already noted in Chapter 1). Refer again to Figure 1, which uses the software FAWAV by J. S. Walker ([WAL]) to illustrate partial sums of the Fourier series for the Dirac mass.

The perceptive reader will have noticed that the Haar series does not give an entirely satisfactory approximation to our function f_N, just because the partial sums each have mean-value zero (which f_N most certainly does not!). Matters are easily remedied by using the decomposition

$$L^2 = V_0 \oplus \bigoplus_0^\infty W_j \qquad\qquad (7.4.1.5)$$

instead of the decomposition

$$L^2 = \bigoplus_{-\infty}^\infty W_j$$

that we have been using. For, with (7.4.1.5), V_0 takes care of the coarse scale behavior all at once, and also gets the mean-value condition right.

Thus we see, in the context of a very simple example, that the partial sums of the Haar series for a function that closely approximates the Dirac mass at the origin give a more accurate and satisfying approximation to the function than do the partial sums of the Fourier series. To be sure, the partial sums of the Fourier series of each f_N tend to that f_N, but the oscillating error persists no matter how high the degree of the partial sum. The situation would be similar if we endeavored to approximate f_N by its Fourier transform.

We close this discussion with some explicit calculations to recap the point that has just been made. It is easy to calculate that the jth Fourier coefficient of the function f_N is

$$\widehat{f_N}(j) = \frac{i 2^{N-1}}{j\pi}\left(e^{-ij/2^N} - 1\right).$$

Therefore, with S_M denoting the Mth partial sum of the Fourier series,

$$\|f_N - S_M\|_{L^2}^2 = \sum_{|j|>M}\left(\frac{2^{N-1}}{j\pi}\right)^2 |e^{-ij/2^N} - 1|^2.$$

Imitating the proof of the integral test for convergence of series, it is now straightforward to see that

$$\| f_N - S_M \|_{L^2}^2 \approx \frac{C}{M}.$$

In short, $\| f_N - S_M \|_{L^2} \to 0$, as $M \to \infty$, at a rate comparable to $M^{-1/2}$, and that is quite slow.

By contrast, if we let $H_M \equiv \sum_{|j| \leq M} 2^{j/2} \mu_j$ then, for $M \geq N - 1$, our earlier calculations show that

$$\| f_N - H_M \|_{L^2}^2 = \sum_{j=-\infty}^{-M-1} 2^j = 2^{-M}.$$

Therefore $\| f_N - H_M \|_{L^2} \to 0$, as $M \to \infty$, at a rate comparable to $2^{-M/2}$, or *exponentially fast*. This is a strong improvement over the convergence supplied by classical Fourier analysis. $\square$

Our next example shows quite specifically that Haar series can beat Fourier series at their own game. Specifically, we shall approximate the function $g(x) \equiv [\cos \pi x] \cdot \chi_{[0,1]}(x)$ both by Haar series and by using the Fourier transform. The Haar series will win by a considerable margin. [**Note:** A word of explanation is in order here. Instead of the function g, we could consider $h(x) \equiv [\cos \pi x] \cdot \chi_{[0,2]}(x)$. Of course the interval $[0, 2]$ is the natural support for a period of the trigonometric function $\cos \pi x$, and the (suitably scaled) *Fourier series* of this function h is just the single term $\cos \pi x$. In this special circumstance Fourier series is hands down the best method of approximation—just because the support of the function is a good fit to the function. Such a situation is too artificial, and not a good test of the method. A more realistic situation is to chop off the cosine function so that its support does not mesh naturally with the period of cosine. That is what the function g does. We give Fourier every possible chance: by approximating with the Fourier *transform*, we allow all possible frequencies, and let Fourier analysis pick those that will best do the job.]

Example 7.4.2 Consider $g(x) = [\cos \pi x] \cdot \chi_{[0,1]}(x)$ as a function on the entire real line. We shall compare and contrast the approximation of g by partial sums using the Haar basis with the approximation of g by "partial sums" of the Fourier transform. Much of what we do here will be traditional hand work; but, at propitious moments, we shall bring the computer to our aid.

Let us begin by looking at the Fourier transform of g. We calculate that

$$
\begin{aligned}
\widehat{g}(\xi) &= \frac{1}{2} \int_0^1 \left(e^{i\pi x} + e^{-i\pi x}\right) e^{ix \cdot \xi} \, dx \\
&= \frac{1}{2} \left[\frac{-e^{i\xi} - 1}{i(\xi + \pi)} + \frac{-e^{i\xi} - 1}{i(\xi - \pi)} \right] \\
&= \frac{-e^{i\xi} - 1}{i(\xi^2 - \pi^2)} \cdot \xi.
\end{aligned}
$$

Observe that the function $\widehat{g}$ is continuous on all of $\mathbb{R}$ and vanishes at ∞. The Fourier inversion formula (Theorem 2.3.13) then tells us that g may be recovered from $\widehat{g}$ by the integral

$$
\frac{1}{2\pi} \int_{\mathbb{R}} \widehat{g}(\xi) e^{-ix \cdot \xi} \, d\xi.
$$

In Chapter 2, we used Gauss-Weierstrass summation in order to effectively implement the idea of summation. However, it is more in the spirit of the present discussion (and also computationally easier) to consider the limit of the integrals

$$
\eta_N(x) \equiv \frac{1}{2\pi} \int_{-N}^{N} \widehat{g}(\xi) e^{-ix \cdot \xi} \, d\xi \tag{7.4.2.1}
$$

as $N \to +\infty$. Elementary calculations show that (7.4.2.1) equals

$$
\begin{aligned}
\eta_N(x) &= \frac{1}{2\pi} \int_{-N}^{N} \int_{-\infty}^{\infty} g(t) e^{i\xi t} \, dt \, e^{-ix\xi} \, d\xi \\
&= \frac{1}{2\pi} \int_0^1 g(t) \int_{-N}^{N} e^{i(t-x)\xi} \, d\xi \, dt
\end{aligned}
$$

$$= \frac{1}{2\pi i} \int_0^1 g(t) \frac{1}{t-x} e^{i\xi(t-x)} \Big]_{-N}^{N} dt$$

$$= \frac{1}{2\pi i} \int_0^1 g(t) \frac{1}{t-x} \left[e^{iN(t-x)} - e^{i(-N)(t-x)} \right] dt$$

$$= \frac{1}{2\pi i} \int_0^1 g(t) \frac{1}{t-x} 2i \sin N(t-x) \, dt$$

$$= \frac{1}{\pi} \int_0^1 g(t) \frac{\sin N(x-t)}{x-t} \, dt$$

$$= \frac{1}{\pi} \int_0^1 \cos \pi t \frac{\sin N(x-t)}{x-t} \, dt. \qquad (7.4.2.2)$$

We see, by inspection of (7.4.2.2), that η_N is a continuous, indeed an analytic, function. Thus it is supported on the entire real line (not on any compact set). Notice further that it could not be the case that $\eta_N = \mathcal{O}(|x|^{-r})$ for some $r > 1$; if it were, then η_N would be in $L^1(\mathbb{R})$ and then $\widehat{\eta_N}$ would be continuous (which it is certainly not). It turns out (we omit the details) that in fact $\eta_N = \mathcal{O}(|x|^{-1})$. This statement says, in a quantitative way, that η_N has a tail.

We can rewrite formula (7.4.2.2) (the last item in our long calculation) in the form

$$\eta_N(x) = \frac{1}{\pi} \int_{\mathbb{R}} g(t) \widetilde{D}_N(x-t) \, dt,$$

where

$$\widetilde{D}_N(t) = \frac{\sin Nt}{\pi t}.$$

The astute reader will realize that the kernel $\widetilde{D}_N$ is quite similar to the Dirichlet kernel that we studied in Chapter 1 in connection with Fourier series. A proof analogous to ones we considered there will show that $\eta_N(x) \to g(x)$ pointwise as $N \to \infty$.

Our calculations confirm that the Fourier transform of g can be "Fourier-inverted" (in the L^2 sense) back to g. But they also show that, for any particular $N > 0$ large, the expression

$$\eta_N(x) \equiv \frac{1}{2\pi} \int_{-N}^{N} \widehat{g}(\xi) e^{-ix\cdot\xi} \, d\xi \qquad (7.4.2.3)$$

is supported *on the entire real line*. [Of course this must be so since, if we replace the real variable x with a complex variable z, then (7.4.2.3) defines an entire function.] Thus, for practical applications, the convergence of η_N to g on the support $[0, 1]$ of g is seriously offset by the fact that η_N has a "tail" that persists no matter how large N is. And the key fact is that the tail is *not small*. This feature is built in just because the function we are expanding has discontinuities.

We now contrast the preceding calculation of the Fourier transform of the function $g(x) = [\cos \pi x] \cdot \chi_{[0,1]}(x)$ with the analogous calculation using the Haar basis (but we shall perform these new calculations with the aid of a computer). The first thing that we will notice is that the only Haar basis elements that end up being used in the expansion of g are *those basis elements that are supported in the interval* $[0, 1]$. For the purposes of signal processing, this is already a dramatic improvement.

Figure 2 shows the Fourier transform approximation to the function g. Specifically, this is a graph of η_{64} created with Walker's software FAWAV. Figure 3 shows the improved approximation attained by η_{128}. Figures 4 and 5, respectively, superimpose the approximations η_{64} and η_{128} against the graph of g. Notice that, while the approximations are reasonable *inside*—and away from the endpoints of—the unit interval, the "inverse" of the Fourier transform goes out of control as x moves left across 0 or as x moves right across 1. By contrast, the Haar series for g (Figure 6, which shows the 129-term Haar series partial sum) is quite tame and gives a good approximation. Figure 7 shows the 257-term Haar series approximation—an even more dramatic improvement.

More precisely, the Haar series partial sums are supported on $[0, 1]$ (just like the function g) and they converge uniformly on $[0, 1)$ to g (exercise). Of course the Haar series is not the final solution either. It has good quantitative behavior, but its qualitative behavior is poor be-

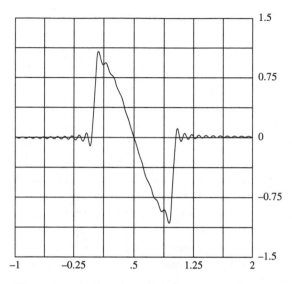

Figure 2. The Fourier transform approximation to g.

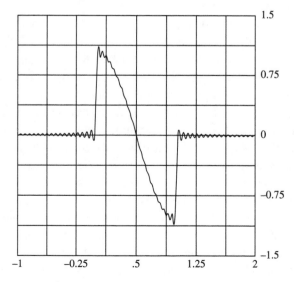

Figure 3. Improved approximation to g.

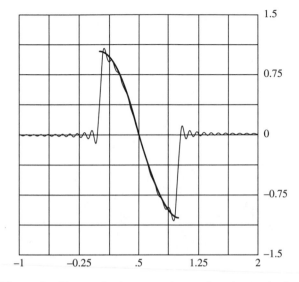

Figure 4.   The graph of η_{64} superimposed on the graph of g.

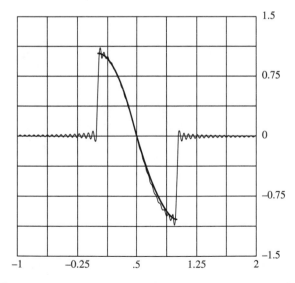

Figure 5.   The graph of η_{128} superimposed on the graph of g.

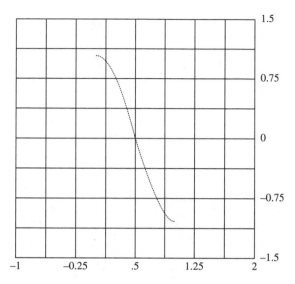

Figure 6. The 129-term Haar series approximation to g.

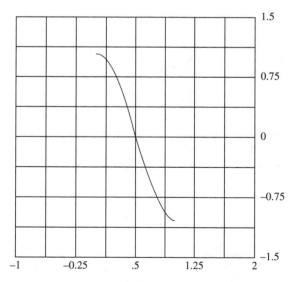

Figure 7. The 257-term Haar series approximation to g.

cause the partial sums are piecewise constant (i.e., *jagged*) functions. We thus begin to see the desirability of smooth wavelets. □

Part of the reason that wavelet sums exhibit this dramatic improvement over Fourier sums is that wavelets provide an "unconditional basis" for many standard function spaces (see [HERG, p. 233 ff.], as well as the discussion in the next section, for more on this idea). Briefly, the advantage that wavelets offer is that we can select only those wavelet basis functions whose supports overlap with the support of the function being approximated. This procedure corresponds, roughly speaking, with the operation of rearranging a series; such rearrangement is possible for series formed from an unconditional basis, but not (in general) with Fourier series.

7.5   Construction of a Wavelet Basis

There exist examples of an MRA for which the scaling function ϕ is smooth and compactly supported. It is known—for reasons connected with the uncertainty principle (Section 2.4)—that there do not exist C^∞ (infinitely differentiable) scaling functions that are compactly supported, or that satisfy the weaker condition that they decay exponentially at infinity—see [HERG, p. 197] for a proof. But there *do exist* compactly supported C^k (k times continuously differentiable) scaling functions for each k. In this section we will give an indication of I. Daubechies's construction of such scaling functions. We begin, however, by first describing the properties of such a scaling function, and how the function might be utilized.

So suppose that ϕ is a scaling function that is compactly supported and is C^k. By the axioms of an MRA, the functions $\{\tau_k \phi\}_{k \in \mathbb{Z}}$ form a basis for V_0. It follows then that the functions $\{\sqrt{2}\alpha_2 \tau_k \phi\}_{k \in \mathbb{Z}}$ form an orthonormal basis for V_1. Written more explicitly, these functions have the form $\sqrt{2}\phi(2x - k)$, and they span V_1. Since $\phi \in V_0 \subseteq V_1$, we may expand ϕ itself in terms of the functions $\sqrt{2}\phi(2x - k)$. Thus

$$\phi(x) = \sum_k c_k \sqrt{2}\phi(2x - k), \qquad (7.5.1)$$

where

$$c_k = \int \phi(x)\sqrt{2}\,\overline{\phi(2x - k)}\,dx.$$

If we set

$$\psi(x) = \sum_k (-1)^k c_{1-k}\sqrt{2}\phi(2x - k), \qquad (7.5.2)$$

then the functions $\psi(x - \ell)$, $\ell \in \mathbb{Z}$, will be orthogonal and will span W_0. To see the first assertion, we calculate the integral

$$\int \psi(x - k)\overline{\psi(x - \ell)}\,dx$$

$$= \int \left[\sum_k (-1)^k c_{1-k}\sqrt{2}\phi(2x - k)\right]$$

$$\times \left[\sum_\ell (-1)^\ell \overline{c_{1-\ell}\sqrt{2}\phi(2x - \ell)}\right] dx$$

$$= \sum_{k,\ell} 2c_{1-k}\overline{c_{1-\ell}}(-1)^{k+\ell} \int \phi(2x - k) \cdot \overline{\phi(2x - \ell)}\,dx.$$

Of course the kth integral in this last sum will be zero if $k \neq \ell$ because of Axiom 5 of an MRA. If instead $k = \ell$, then the integral evaluates to $1/2$ by a simple change of variable. If we mandate in advance that $\int |\phi|^2 = 1$, then $\sum_k |c_k|^2 = 1$ and the result follows.

As for the functions $\psi(x - \ell)$ spanning W_0, it is slightly more convenient to verify that $\{\phi(x - m)\}_{m \in \mathbb{Z}} \cup \{\psi(x - \ell)\}_{\ell \in \mathbb{Z}}$ spans $V_1 = V_0 \oplus W_0$. Since the functions $\phi(2x - n)$ already span V_1, it is enough to express each of them as a linear combination of functions $\phi(x - m)$ and $\psi(x - \ell)$. If this is to be so, then the coefficient $a_n(m)$ of $\phi(2x - m)$

will have to be

$$
a_n(m) = \int \phi(2x - n)\overline{\phi(x - m)}\, dx
$$

$$
= \int \phi(2x - n) \sum_k \overline{c_k} \sqrt{2}\, \overline{\phi(2x - 2m - k)}\, dx
$$

$$
= \sum_k \sqrt{2}\, \overline{c_k} \int \phi(2x - n)\overline{\phi(2x - 2m - k)}\, dx.
$$

The summand can be nonzero only when $n = 2m + k$, that is, when $k = n - 2m$. Hence

$$
a_n(m) = \frac{1}{\sqrt{2}}\, \overline{c_{n-2m}}.
$$

Likewise, the coefficient $b_n(m)$ of $\psi(2x - \ell)$ will have to be

$$
b_n(m) = \int \phi(2x - n)\overline{\psi(x - \ell)}\, dx
$$

$$
= \int \phi(2x - n) \sum_k (-1)^k \overline{c_{1-k}} \sqrt{2}\, \overline{\phi(2x - 2\ell - k)}\, dx
$$

$$
= \sum_k \sqrt{2}\, \overline{c_{1-k}} \int \phi(2x - n)\overline{\phi(2x - 2\ell - k)}\, dx.
$$

Of course this integral can be nonzero only when $2\ell + k = n$, that is, when $k = n - 2\ell$. Thus the ℓth coefficient is

$$
b_n(m) = \frac{(-1)^n}{\sqrt{2}}\, \overline{c_{1-n+2\ell}}.
$$

Thus our task reduces to showing that

$$
\phi(2x - n) = \frac{1}{\sqrt{2}} \left[\sum_{m \in \mathbb{Z}} \overline{c_{n-2m}}\phi(x - m) \right.
$$

$$
\left. + (-1)^n \sum_{\ell \in \mathbb{Z}} \overline{c_{1-n+2\ell}}\psi(x - \ell) \right].
$$

If we plug equations (7.5.1) and (7.5.2) into this last equation, we end up with an identity in the functions $\phi(2x - p)$, which in turn reduces to an algebraic identity on the coefficients. This algebraic lemma is proved in [STR3, p. 546]; it is similar in spirit to the calculations that precede Theorem 7.3.1. We shall not provide the details here.

The Haar basis, while elementary and convenient, has several shortcomings. Chief among these is the fact that each basis element is discontinuous. One consequence is that the Haar basis does a poor job of approximating continuous functions. A more profound corollary of the discontinuity is that the Fourier transform of a Haar wavelet dies like $1/x$ at infinity, and hence is not integrable. It is desirable to have smooth wavelets, for as we know (Chapter 2), the Fourier transform of a smooth function dies rapidly at infinity. The Daubechies wavelets are important partly because they are as smooth as we wish; for a thorough discussion of these see [WAL] or [HERG].

Our discussion of the Haar wavelets (or MRA) already captures the spirit of wavelet analysis. In particular, it generates a complete orthonormal basis for L^2 with the property that finite sums of the basis elements give a good approximation (better than partial sums of Fourier series exponentials) to the Dirac mass δ. Since any L^2 function f can be written as $f = f * \delta$, it follows (subject to checking that Haar wavelets interact nicely with convolution) that any L^2 function with suitable properties will have a good approximation by wavelet partial sums.

The Haar wavelets are particularly effective at encoding information coming from a function that is constant on large intervals. The reason is that the function ψ integrates to zero—we say that it has "mean value zero." Thus integration against ψ annihilates constants. If we want a wavelet that compresses more general classes of functions, then it is natural to mandate that the wavelet annihilate first linear functions, then quadratic functions, and so forth. In other words, we typically demand that our wavelet satisfy

$$\int \psi(x)x^j \, dx = 0 \, , \qquad j = 0, 1, 2, \ldots, L - 1 \qquad (7.5.3)$$

for some prespecified positive integer L. In this circumstance we say that ψ has "L vanishing moments." Of course, it would be helpful, although it is not necessary, in achieving these vanishing moment conditions to have a wavelet ψ that is smooth.

It is a basic fact that smooth wavelets must have vanishing moments. More precisely, if ψ is a C^k wavelet such that

$$|\psi(x)| \leq C(1 + |x|)^{-k-2},$$

then it must be that

$$\int x^j \psi(x)\, dx = 0, \qquad 0 \leq j \leq k.$$

Here is a sketch of the reason:

Let $\{\psi_k^j\}_{k \in \mathbb{Z}}$ be the basis generated by ψ for the space W_j. Let $j \gg j'$. Then ψ_k^j lives on a *much smaller scale* than does $\psi_{k'}^{j'}$. Therefore $\psi_{k'}^{j'}$ is (essentially) a Taylor polynomial on the interval where ψ_k^j lives. Hence the orthogonality

$$\int \psi_k^j \overline{\psi_{k'}^{j'}} = 0$$

is essentially equivalent to

$$\int \psi_k^j(x) \overline{x^m}\, dx = 0$$

for appropriate m. This is the vanishing-moment condition.

It is a basic fact about calculus that a function with many vanishing moments must oscillate a great deal. For instance, if a function f is to integrate to 0 against both 1 and x, then f integrates to zero against all linear functions. So f itself cannot be linear; it must be at least quadratic. That gives one oscillation. Likewise, if f is to integrate to 0 against $1, x, x^2$, then f must be at least cubic. That gives two oscillations. And so forth.

Remark. It is appropriate at this point to offer an aside about why there cannot exist a C^∞ wavelet with compact support. First, a C^∞ wavelet ψ must have vanishing moments of all orders. Passing to the Fourier transform (and using Propositions 2.1.2 and 2.1.3), we see therefore that $\widehat{\psi}$ vanishes to infinite order at the origin (i.e., $\widehat{\psi}$ and all its derivatives vanish at 0). If ψ were compactly supported, then $\widehat{\psi}$ would be analytic (see [KRA3, §2.4]); the infinite-order vanishing then forces $\widehat{\psi}$, and hence ψ, to be identically zero.

A Combinatorial Construction of the Daubechies Wavelets

With these thoughts in mind, let us give the steps that explain how to use Daubechies's construction to create a *continuous* wavelet. [Constructing a C^1 or smoother wavelet follows the same lines, but is much more complicated.] We begin by nearly repeating the calculations at the beginning of this section, but then we add a twist.

Imagine functions ϕ and ψ, both continuous, and satisfying

$$\int_{-\infty}^{+\infty} \phi(x)\,dx = 1 \;, \quad \int_{-\infty}^{+\infty} |\phi(x)|^2\,dx = 1 \;, \quad \int_{-\infty}^{+\infty} |\psi(x)|^2\,dx = 1.$$

$$(7.5.4)$$

We know from the MRA axioms that the function ϕ must generate the basic space V_0. Moreover, we require that $V_0 \subseteq V_1$. It follows that

$$\phi(x) = \sum_{j \in \mathbb{Z}} c_j \sqrt{2} \phi(2x - j) \tag{7.5.5}$$

for some constants c_j. The equation

$$\psi(x) = \sum_{j \in \mathbb{Z}} (-1)^j c_{1-j} \sqrt{2} \phi(2x - j) \tag{7.5.6}$$

defines a wavelet ψ such that $\{\tau_k \psi\}$ spans the subspace W_0. Notice that equations (7.5.5) and (7.5.6) generalize the relations that we had between ϕ and ψ for the Haar basis.

Equation (7.5.5), together with the first two integrals in equation (7.5.4), shows that

$$\sum_{j\in\mathbb{Z}} c_j = \sqrt{2} \quad \text{and} \quad \sum_{j\in\mathbb{Z}} |c_j|^2 = 1. \qquad (7.5.7)$$

If, for specificity, we take $L = 2$, equation (7.5.3) combined with equation (7.5.6) imply that

$$\sum_{j\in\mathbb{Z}} (-1)^j c_j = 0 \quad \text{and} \quad \sum_{j\in\mathbb{Z}} j(-1)^j c_j = 0. \qquad (7.5.8)$$

In fact one can solve the equations in (7.5.7) and (7.5.8); one standard solution is

$$c_0 = \frac{1+\sqrt{3}}{4\sqrt{2}} , \ c_1 = \frac{3+\sqrt{3}}{4\sqrt{2}} , \ c_2 = \frac{3-\sqrt{3}}{4\sqrt{2}} , \ c_3 = \frac{1-\sqrt{3}}{4\sqrt{2}} ,$$

$$(7.5.9)$$

and all other $c_j = 0$.

Now here comes the payoff. Using these values of c_j, we may define

$$\phi_0(x) = \chi_{[0,1)}(x)$$
$$\phi_j(x) = \sum_{\ell\in\mathbb{Z}} c_\ell \sqrt{2}\phi_{j-1}(2x - \ell) \quad \text{when } j \geq 1.$$

The functions ϕ_j, iteratively defined as above, converge to a continuous function ϕ that is supported in $[0, 3]$. It can then be seen from equation (7.5.6) that the corresponding function ψ is continuous and supported in $[-1, 2]$.

Now that suitable ϕ and ψ have been found, we may proceed step-by-step as we did with the construction of the Haar wavelet basis. We shall not provide the details, but instead refer the reader to [WAL], from which this particular presentation of wavelet ideas derives.

The Daubechies Wavelets from the Point of View of Fourier Analysis

We now give a last loving look at the Daubechies wavelet construction—this time from the point of view of Fourier analysis. Observe that if ϕ is smooth and of compact support, then the sum in equation (7.5.5) must be finite. Using Proposition 2.2.4, we may calculate that

$$\left(\sqrt{2}\phi(2\cdot -j)\right)\widehat{\ }(\xi) = \left(\sqrt{2}\alpha_2\tau_j\phi\right)\widehat{\ }(\xi)$$

$$= \frac{\sqrt{2}}{2}\left(\tau_j\phi\right)\widehat{\ }(\xi/2) = \frac{1}{\sqrt{2}}e^{ij\xi/2}\widehat{\phi}(\xi/2).$$

If we set

$$m(\xi) = \frac{1}{\sqrt{2}}\sum_j c_j e^{ij\xi},$$

where the c_j are as in equation (7.5.5), then we may write (applying the Fourier transform to the sum in (7.5.5))

$$\widehat{\phi}(\xi) = m(\xi/2)\widehat{\phi}(\xi/2). \tag{7.5.10}$$

We call the function m a *low-pass filter*.

Iterating this last identity yields

$$\widehat{\phi}(\xi) = m(\xi/2)m(\xi/4)\widehat{\phi}(\xi/4)$$

$$\cdots$$

$$\widehat{\phi}(\xi) = m(\xi/2)m(\xi/4)\cdots m(\xi/2^p)\widehat{\phi}(\xi/2^p).$$

Since $\widehat{\phi}(0) = \int \phi(x)\,dx = 1$, we find in the limit that

$$\widehat{\phi}(\xi) = \prod_{p=1}^{\infty} m(\xi/2^p). \tag{7.5.11}$$

Now the orthonormality of the $\{\phi(x-j)\}$ implies the identity

$$|m(\xi)|^2 + |m(\xi+\pi)|^2 \equiv 1. \tag{7.5.12}$$

To wit, we calculate using Plancherel's theorem (and with $\delta_{j,k}$ denoting the Kronecker delta) that

$$\delta_{j,0} = \int_{\mathbb{R}} \phi(x)\overline{\phi(x-j)}\,dx$$

$$= \frac{1}{2\pi} \int_{\mathbb{R}} |\widehat{\phi}(\xi)|^2 e^{-ij\xi}\,d\xi$$

$$= \frac{1}{2\pi} \sum_{\ell=-\infty}^{\infty} \int_{2\ell\pi}^{2(\ell+1)\pi} |\widehat{\phi}(\xi)|^2 e^{-ij\xi}\,d\xi$$

$$= \frac{1}{2\pi} \sum_{\ell=-\infty}^{\infty} \int_0^{2\pi} |\widehat{\phi}(\lambda + 2\ell\pi)|^2 e^{-ij\mu}\,d\lambda$$

$$= \frac{1}{2\pi} \int_0^{2\pi} \left(\sum_{\ell \in \mathbb{Z}} |\widehat{\phi}(\lambda + 2\ell\pi)|^2 \right) e^{-ij\lambda}\,d\lambda.$$

This calculation tells us that the 2π-periodic function $\sum_\ell |\widehat{\phi}(\mu + 2\ell\pi)|^2$ has Fourier coefficient 1 at the frequency 0 and all other Fourier coefficients zero. In other words,

$$\sum_\ell |\widehat{\phi}(\mu + 2\ell\pi)|^2 = 1 \qquad \text{a.e.}$$

Using (7.5.10), we find that

$$\sum_\ell |\widehat{\phi}(\xi + \ell\pi)|^2 |m(\xi + \ell\pi)|^2 = 1 \qquad \text{a.e.}$$

Now separating into sums over even and odd indices ℓ, and using the 2π-periodicity of m, yields (7.5.12).

Running our last arguments backwards, it can be shown that if m is a trigonometric polynomial satisfying (7.5.12) and such that $m(0) = 1$, then the product in (7.5.11) converges uniformly on compact sets to a function $\widehat{\phi} \in L^2$. Also, if $\widehat{\phi}$ decays sufficiently rapidly ($|\widehat{\phi}| = \mathcal{O}(1 + |\xi|)^{-1/2-\epsilon}$ will do), then its inverse Fourier transform ϕ is the scaling function of an MRA.

In summary, if we can find a trigonometric polynomial m satisfying (7.5.12) with $m(0) = 1$ and so that the resulting $\widehat{\phi}$ satisfies

$|\widehat{\phi}(\xi)| = \mathcal{O}(1 + |\xi|)^{-k-1-\epsilon}$ for some integer $k \geq 0$, then ϕ will be a compactly supported wavelet of class C^k. It should be noted that *finding such a trigonometric polynomial m is hard work*.

As a final note, if ϕ_0 is a "nice" function with $\widehat{\phi}_0(0) = 1$ and if we define ϕ_K inductively by

$$\widehat{\phi_K}(\xi) = m(\xi/2)\widehat{\phi_{K-1}}(\xi/2), \qquad (7.5.13)$$

then, by (7.5.11),

$$\lim_{K \to \infty} \widehat{\phi_K}(\xi) = \lim_{K \to \infty} \widehat{\phi_0}(\xi/2^K) \prod_{k=1}^{K} m(\xi/2^k) \equiv \widehat{\phi}(\xi).$$

This last equation *defines* $\widehat{\phi}$, and therefore ϕ itself. Unraveling the Fourier transform in (7.5.13), we conclude that $\phi = \lim \phi_K$, where ϕ_K is defined inductively by

$$\phi_K(x) = \sum_j c_j \sqrt{2}\phi_{K-1}(2x - j).$$

The analysis that we have just given shows that wavelet theory is firmly founded on invariance properties of the Fourier transform. In other words, wavelet theory does not displace the classical Fourier theory; rather, it builds on those venerable ideas.

Reflective Remarks

The iterative procedure that we have used to construct the scaling function ϕ has some interesting side effects. One is that ϕ has certain self-similarity properties that are reminiscent of fractals.

We summarize the very sketchy presentation of the present chapter by pointing out that an MRA (and its generalizations to wavelet packets and to the local cosine bases of Coifman and Meyer [HERG]) gives a "designer" version of Fourier analysis that retains many of the favorable features of classical Fourier analysis, but also allows the user to adapt the system to problems at hand. We have given a construction that is particularly well adapted to detecting spikes in a sound wave, and therefore is useful for denoising. Other wavelet constructions have proved useful in signal compression, image compression, and other engineering applications.

We now present two noteworthy mathematical applications of wavelet theory. They have independent interest, but are also closely connected to each other (by way of wavelet theory) and to ideas in the rest of the book. We shall indicate some of these connections. One of these applications is to see that wavelets give a natural unconditional basis for many of the classical Banach spaces of analysis. The other is to see that a Calderón-Zygmund operator is essentially diagonal when expressed as a bi-infinite matrix with respect to a wavelet basis.

We sketch some of the ideas adherent to the previous paragraph, and refer the reader to [DAU] and [MEY3] for the details.

Wavelets as an Unconditional Basis

Recall that a set of vectors $\{e_0, e_1, \ldots\}$ in a Banach space X is called an *unconditional basis* if it has the following properties:

(7.5.14) For each $x \in X$ there is a unique sequence of scalars $\alpha_0, \alpha_1, \ldots$ such that

$$x = \sum_{j=0}^{\infty} \alpha_j e_j,$$

in the sense that the partial sums $S_N \equiv \sum_{j=0}^{N} \alpha_j e_j$ converge to x in the topology of the Banach space.

(7.5.15) There exists a constant C such that, for each integer m, for each sequence $\alpha_0, \alpha_1, \ldots$ of coefficients as in (7.5.14), and for any sequence $\beta_0, \beta_1, \ldots$ satisfying $|\beta_k| \leq |\alpha_k|$ for all $0 \leq k \leq m$, we have

$$\left\| \sum_{k=0}^{m} \beta_k e_k \right\| \leq C \cdot \left\| \sum_{k=0}^{m} \alpha_k e_k \right\|. \qquad (7.5.15.1)$$

We commonly describe Property (7.5.14) with the phrase "$\{e_j\}$ is a *Schauder basis* for X." The practical significance of Property (7.5.15) is that we can decide whether a given formal series $\beta_0 e_0 + \beta_1 e_1 + \cdots$ converges to an element $y \in X$ simply by checking the sizes of the coefficients.

Let us consider the classical L^p spaces on the real line. We know by construction that the wavelets form an orthonormal basis for $L^2(\mathbb{R})$. In particular, the partial sums are dense in L^2. So they are also dense in $L^2 \cap L^p$, $2 < p < \infty$, in the L^2 topology. It follows that they are dense in L^p in the L^p topology for this range of p, since the L^p norm then dominates the L^2 norm (the argument for $p < 2$ involves some extra tricks that we omit). Modulo some technical details, this says in effect that the wavelets form a Schauder basis for L^p, $1 < p < \infty$. Now let us address the "unconditional" aspect.

It can be shown that Property (7.5.15) holds for all sequences $\{\beta_j\}$ if and only if it holds in all the special cases $\beta_j = \pm\alpha_j$. Suppose that

$$L^p \ni f = \sum_{j,k} \alpha_{j,k} \psi_k^j;$$

then of course it must be that $\alpha_{j,k} = \int f(x)\overline{\psi_k^j(x)}\,dx = \langle f, \psi_k^j \rangle$ (by the orthonormality of the wavelets). So we need to show that, for any choice of $\mathbf{w} \equiv \{w_{j,k}\} = \{\pm 1\}$, the operator $T_\mathbf{w}$ defined by

$$T_\mathbf{w} f = \sum_{j,k} w_{j,k} \langle f, \psi_k^j \rangle \psi_k^j$$

is a bounded operator on L^p. We certainly know that $T_\mathbf{w}$ is bounded on L^2, for

$$\|T_\mathbf{w} f\|_{L^2}^2 = \sum_{j,k} |w_{j,k}\langle f, \psi_k^j \rangle|^2 = \sum_{j,k} |\langle f, \psi_k^j \rangle|^2 = \|f\|_{L^2}^2.$$

The L^p boundedness will then follow from the Calderón-Zygmund theorem (see Theorem 6.2.10) provided that we can prove suitable estimates for the integral kernel of $T_\mathbf{w}$. The necessary estimates are these (which should look familiar):

$$|k(x, y)| \leq \frac{C}{|x - y|}$$

and

$$\left|\frac{\partial}{\partial x}k(x, y)\right| + \left|\frac{\partial}{\partial y}k(x, y)\right| \leq \frac{C}{|x - y|^2}.$$

These are proved in Lemma 9.1.5 on p. 296 of [DAU]. We shall not provide the details here. □

Wavelets and Almost Diagonalizability

Now let us say something about the "almost diagonalizability" of Calderón-Zygmund operators with respect to a wavelet basis. In fact we have already seen an instance of this phenomenon: the kernel K for the operator $T_{\mathbf{w}}$ that we just considered must have the form

$$K(x, y) = \sum_{j,k} w_{j,k} \psi_k^j(x)\overline{\psi_k^j(y)}. \tag{7.5.16}$$

Do not be confused by the double indexing! If we replace the double index (j, k) by the single index ℓ, then the kernel becomes

$$K(x, y) = \sum_{\ell} w_{\ell} \psi_{\ell}(x)\overline{\psi_{\ell}(y)}.$$

We see that the kernel, an instance of a singular integral kernel, is plainly diagonal.

In fact it is easy to see that an operator given by a kernel that is diagonal with respect to a wavelet basis induces an operator that is bounded on L^p. For let

$$K(x, y) = \sum_{j,k} \alpha_{j,k} \psi_k^j(x)\overline{\psi_k^j(y)}.$$

Then the operator

$$T_K : f \mapsto \int K(x, y) f(y) \, dy$$

satisfies (at least at a computational level)

$$T_K f(x) = \sum_{j,k} \alpha_{j,k} \langle f, \psi_k^j \rangle \psi_k^j (x).$$

Since $\{\psi_k^j\}$ forms an orthonormal basis for L^2, we see that if each $\alpha_{j,k} = 1$, then the last line is precisely f. If instead the $\alpha_{j,k}$ form a bounded sequence, then the last displayed line represents a bounded operator on L^2. In fact it turns out that T_K must be (essentially) the sort of operator that is being described in the $T(1)$ theorem (Section 6.4). A few details follow:

A translation-invariant operator T, with kernel k, is a Calderón-Zygmund operator (in the generalized sense of Theorem 6.4.6) if and only if it is "essentially diagonal" with respect to a wavelet basis, in the sense that the matrix entries die off rapidly away from the diagonal of the matrix. To see the "only if" part of this assertion, one calculates

$$\langle T \psi_k^j, \psi_{k'}^{j'} \rangle = \int_{I'} \int_I k(x - y) \psi_k^j (x) \psi_{k'}^{j'} (y) \, dx dy,$$

where I is the interval that is the support of ψ and I' is the interval that is the support of ψ'. One then exploits the mean-value-zero properties of ψ_k^j and $\psi_{k'}^{j'}$, together with the estimates

$$|k(x, y)| \leq \frac{C}{|x - y|}$$

and

$$\left| \frac{\partial}{\partial x} k(x, y) \right| + \left| \frac{\partial}{\partial y} k(x, y) \right| \leq \frac{C}{|x - y|^2}.$$

After some calculation, the result is that

$$\left| \langle T \psi_k^j, \psi_{k'}^{j'} \rangle \right| \leq C \cdot e^{-c \cdot \rho(\zeta, \zeta')},$$

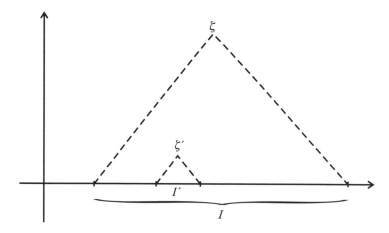

Figure 1. The geometry of the hyperbolic plane.

where ρ is the hyperbolic (Poincaré) distance between the points ζ, ζ' that are associated with I, I' as in Figure 1. We shall not provide the proofs of these assertions, but instead refer the reader to [MEY1], [MEY3], and [DAU].

Exercise. Calculate the matrix of the Hilbert transform

$$f \longmapsto \text{P.V.} \int \frac{f(t)}{x - t}\, dt$$

with respect to the Haar basis. Conclude that the Hilbert transform is bounded on $L^2(\mathbb{R})$.

In effect, wavelet analysis has caused harmonic analysis to reinvent itself. Wavelets and their generalizations are a powerful new tool that allow localization in both the space and phase variables. They are useful in producing unconditional bases for classical Banach spaces. They also provide flexible methods for analyzing integral operators. The subject of wavelets promises to be a fruitful area of investigation for many years to come.

CHAPTER 8

A Retrospective

8.1   Fourier Analysis: An Historical Overview

We have seen that the basic questions of Fourier analysis grew out of eighteenth-century studies of the wave equation.

Decades later, J. Fourier found an algorithm for expanding an "arbitrary" function on $[0, 2\pi)$ in terms of sines and cosines. Where Fourier was motivated by physical problems and was perhaps not as mathematically rigorous as we would like, P. G. L. Dirichlet began the process of putting the theory of Fourier series on a rigorous footing.

Fourier analysis has served as a pump for much of modern analysis. G. Cantor's investigations of set theory were motivated in part by questions of sets of uniqueness for Fourier series. H. Lebesgue's measure theory and the twentieth-century development of functional analysis have grown out of questions of Fourier analysis. Fourier analysis— particularly the Fourier transform—and partial differential equations have grown up hand in hand.

In the 1920s, Marcel Riesz proved that the Hilbert transform is bounded on $L^p(\mathbb{T})$, $1 < p < \infty$. In one fell swoop, he established norm convergence for Fourier series, invented the subject of interpolation theory, and firmly planted singular integral operators at the center of harmonic analysis.

The theory of singular integrals took the Hilbert transform and put it in the context of a general theory of operators. A. P. Calderón,

A. Zygmund, E. M. Stein, and others have shown how singular integral theory is the natural venue for studying problems of partial differential equations and of harmonic analysis.

The development of algebras of pseudodifferential operators in the 1960s and 1970s, and of L. Hörmander's more general theory of Fourier integral operators, can be thought of as "next generation" theories of Fourier analysis. They fit singular integrals into powerful algebras of operators that can be used to attack a variety of problems in analysis.

The $T(1)$ Theorem of David-Journé-Semmes assimilates all that we have learned about singular integrals, algebras of operators, and commutators of singular integrals and provides a powerful weapon for analyzing translation noninvariant operators and operators that arise naturally in many settings of geometric analysis.

The new theory of wavelets has given harmonic analysis a way to reinvent itself. Now we can design a Fourier analysis to fit a given problem. The image compression technique currently used by the U.S. Federal Bureau of Investigation to electronically store fingerprints is based on a wavelet algorithm; current methods for restoring recordings of Enrico Caruso and of Johannes Brahms use wavelet signal processing algorithms (and their generalizations involving local cosine bases). The M-bone project, which allows one to view a remote event (such as a lecture at another university) in real time on a work station with a signal received over a telephone line, is possible only because of image compression techniques enabled by wavelet algorithms.

In fact the partial sums that arise naturally in wavelet expansions define Calderón-Zygmund operators, such as the ones we discussed in Chapter 5 (see [HERG, p. 233 ff.] for more on this notion) and also at the end of Chapter 7. Thus we see the newest ideas in the subject fertilizing some of the most venerable ones.

Fourier analysis and harmonic analysis are not just mathematical theories, but are continuing processes. The future of the subject is both vast and unpredictable. Perhaps this book will have helped the reader to begin to appreciate the scope and significance of harmonic analysis.

Appendices and Ancillary Material

Appendix I
The Existence of Testing Functions and
Their Density in L^p

Consider functions on $\mathbb{R}^1$. The real function

$$\phi(x) = \begin{cases} e^{-1/x^2} & \text{if} \quad x > 0 \\ 0 & \text{if} \quad x \leq 0 \end{cases}$$

is infinitely differentiable (C^∞) on the real line (use l'Hôpital's Rule). As a result, the function

$$\psi(x) = \phi(x + 1) \cdot \phi(-x + 1)$$

is C^∞ and is identically zero outside the set $(-1, 1)$. We say that ψ is a "C^∞ function with compact support." The *support* of ψ—that is, the closure of the set on which ψ is nonzero—is the interval $[-1, 1]$. Given any compact interval $[a, b]$, it is clear that this construction may be adapted to produce a C^∞ function whose support is precisely $[a, b]$.

The C^∞ functions with compact support (usually denoted C_c^∞ or $C_c^\infty(\mathbb{R})$ for specificity) form a dense subset of $L^p(\mathbb{R})$, $1 \leq p < \infty$. To see this, first note that the characteristic function of an interval I may be approximated in L^p norm by a C_c^∞ function (see Figure 1). We

315

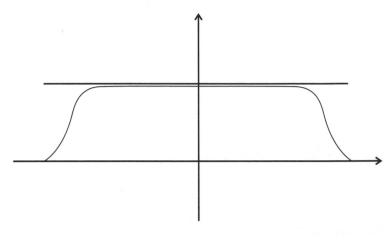

Figure 1.   Approximating a characteristic function by a C_c^∞ function.

achieve that approximation by considering $\psi_\epsilon(x) \equiv \epsilon^{-1}\psi(x/\epsilon)$ and then letting $u_\epsilon = \psi_\epsilon * \chi_I$.

Next, we note that by construction the simple functions are dense in L^p. By the outer regularity of measure, any simple function in L^p may be approximated by a finite sum of characteristic functions of intervals. Each characteristic function of an interval may be approximated in L^p norm by a smooth "bump," as in the last paragraph. That completes a sketch of the proof.

Finally, we use this information to prove that, if $1 \le p < \infty$ and $f \in L^p$, then

$$\lim_{a \to 0} \|f - \tau_a f\|_{L^p} = 0.$$

The assertion is plainly true if f is continuous and compactly supported (i.e., $f \in C_c$), hence uniformly continuous. For any L^p function f, let $\epsilon > 0$ and choose $\phi \in C_c$ such that $\|f - \phi\|_{L^p} < \epsilon$. Then

$$\|f - \tau_a f\|_{L^p} \le \|f - \phi\|_{L^p} + \|\phi - \tau_a \phi\|_{L^p} + \|\tau_a \phi - \tau_a f\|_{L^p}$$

$$= \|f - \phi\|_{L^p} + \|\phi - \tau_a \phi\|_{L^p} + \|\tau_a(\phi - f)\|_{L^p}$$

$$\leq \epsilon + \|\phi - \tau_a \phi\|_{L^p} + \|\phi - f\|_{L^p}$$

$$\leq \epsilon + \|\phi - \tau_a \phi\|_{L^p} + \epsilon.$$

The middle expression tends to zero because ϕ is uniformly continuous (and has compact support). So the result is proved.

We conclude this Appendix by making several remarks. First, all of the constructions presented so far work in higher dimensions, with only small modifications. Second, on the circle group $\mathbb{T}$, the trigonometric polynomials also form a useful dense set in L^p, $1 \leq p < \infty$. This can be seen by imitating the constructions already given, or by using the Stone-Weierstrass theorem (Appendix VIII). Lastly, if (as in the proof of Proposition 2.1.2), one has a differentiable $f \in L^1$ with derivative $\partial f / \partial x_j$ in L^1, then slight modifications of our arguments in this Appendix allow one to approximate such an f by C_c^∞ functions ϕ_k so that $\phi_k \to f$ in the L^1 topology and $\partial \phi_k / \partial x_j \to \partial f / \partial x_j$ in the L^1 topology. Variants of this last remark will be used throughout the book.

Appendix II
Schwartz Functions and the Fourier Transform

A *Schwartz function* ϕ is an infinitely differentiable function such that, for any multi-indices α and β, the expression

$$\rho_{\alpha,\beta}(\phi) \equiv \sup_x \left| x^\alpha \frac{\partial}{\partial x^\beta} \phi(x) \right|$$

is finite. We let $\mathcal{S}$ denote the space of Schwartz functions, and topologize it using the semi-norms $\rho_{\alpha,\beta}$. Because of the facts that $\widehat{f'}(\xi) = -i\xi \widehat{f}(\xi)$ and $[\widehat{f}]'(\xi) = \widehat{ixf}(\xi)$, it is easy to see that the Fourier transform takes Schwartz functions to Schwartz functions. So does the inverse Fourier transform. Since the Fourier transform is one-to-one, it follows that $\widehat{}$ is a bicontinuous isomorphism of $\mathcal{S}$ to itself.

Appendix III
The Interpolation Theorems of Marcinkiewicz and Riesz-Thorin

The simplest example of an interpolation question is as follows. Suppose that the linear operator T is bounded on L^1 and bounded on L^2. Does it follow that T is bounded on L^p for $1 < p < 2$? [The space L^p here is an instance of what is sometimes called an "intermediate space" between L^1 and L^2.] Note that this question is similar to (but not precisely the same as) one that we faced when considering the L^p boundedness of Calderón-Zygmund singular integral operators. Here we record (special) versions of the Riesz-Thorin Theorem (epitomizing the *complex* method of interpolation) and the Marcinkiewicz Interpolation Theorem (epitomizing the *real* method of interpolation) that are adequate for the applications in the present book.

Theorem (Riesz-Thorin): *Let* $1 \leq p_0 < p_1 \leq \infty$. *Let* T *be a linear operator on* $L^{p_0} \cap L^{p_1}$ *such that*

$$\|Tf\|_{L^{p_0}} \leq C_0 \cdot \|f\|_{L^{p_0}}$$

and

$$\|Tf\|_{L^{p_1}} \leq C_1 \cdot \|f\|_{L^{p_1}}.$$

If $0 \leq t \leq 1$ *and*

$$\frac{1}{p} = \frac{1}{p_t} = \frac{1-t}{p_0} + \frac{t}{p_1},$$

then we have

$$\|Tf\|_{L^p} \leq C_0^{1-t} \cdot C_1^t \cdot \|f\|_{L^p}.$$

Recall that, for $1 \leq p < \infty$, we say that a measurable function f is *weak-type p* if there is a constant $C > 0$ such that, for every $\lambda > 0$,

$$m\{x : |f(x)| > \lambda\} \leq \frac{C}{\lambda^p}.$$

We say that f is *weak-type* ∞ if it is just L^∞. A linear operator T is said to be of *weak-type* (p, p) if there is a constant $C > 0$ such that, for each $f \in L^p$ and each $\lambda > 0$,

$$m\{x : |Tf(x)| > \lambda\} \le C \cdot \frac{\|f\|_{L^p}^p}{\lambda^p}.$$

An operator is *weak-type* ∞ if it is simply bounded on L^∞ in the classical sense. An operator is said to be *strong-type* (p, p) (or, more generally, (p, q)) if it is bounded from L^p to L^p (or from L^p to L^q) in the classical sense discussed in Chapter 0.

Now we have

Theorem (Marcinkiewicz): *Let* $1 \le p_0 < p_1 \le \infty$. *If* T *is a (sub-)linear operator on* $L^{p_0} \cap L^{p_1}$ *such that* T *is of weak-type* (p_0, p_0) *and also of weak-type* (p_1, p_1), *then for every* $p_0 < p < p_1$ *we have*

$$\|Tf\|_{L^p} \le C_p \cdot \|f\|_{L^p}.$$

Here the constant C_p *depends on* p *and, in general, will blow up as either* $p \to p_0^+$ *or* $p \to p_1^-$.

These interpolation theorems have been generalized in a number of respects. If T is a (sub-)linear operator that maps L^{p_0} to L^{q_0} and L^{p_1} to L^{q_1} (either weakly or strongly), and if $p_0 < p_1$, $p_0 \le q_0$, $p_1 \le q_1$, then T maps L^{p_t} to L^{q_t} strongly, where

$$\frac{1}{p_t} = (1 - t) \cdot \frac{1}{p_0} + t \cdot \frac{1}{p_1}$$

and

$$\frac{1}{q_t} = (1 - t) \cdot \frac{1}{q_0} + t \cdot \frac{1}{q_1}.$$

We shall occasionally use this more general version of Marcinkiewicz interpolation.

Appendix IV
Hausdorff Measure and Surface Measure

If $\Omega \subseteq \mathbb{R}^N$ has C^1 boundary, then we use the symbol $d\sigma$ to denote $(N - 1)$-dimensional area measure on $\partial\Omega$. This concept is fundamental; we discuss, but do not prove, the equivalence of several definitions for $d\sigma$. A thorough consideration of geometric measures on lower-dimensional sets may be found in the two masterpieces [FE1] and [WHI].

First we consider a version of a construction due to Hausdorff. Let $S \subseteq \mathbb{R}^N$ and $\delta > 0$. Let $\mathcal{U} = \{U_\alpha\}_{\alpha \in A}$ be an open covering of S. Call $\mathcal{U}$ a δ-*admissible covering* if each U_α is an open Euclidean N-ball of radius $0 < r_\alpha < \delta$. If $0 \le k \in \mathbb{Z}$, let M_k be the usual k-dimensional Lebesgue measure of the unit ball in $\mathbb{R}^k$ (e.g., $M_1 = 2$, $M_2 = \pi$, $M_3 = 4\pi/3$, etc.). Define

$$\mathcal{H}_\delta^k(S) = \inf \left\{ \sum_{\alpha \in A} M_k r_\alpha^k : \mathcal{U} = \{U_\alpha\}_{\alpha \in A} \right.$$

$$\text{is a } \delta\text{-admissible cover of } S \}.$$

Clearly, $\mathcal{H}_\delta^k(S) \le \mathcal{H}_{\delta'}^k(S)$ if $0 < \delta' < \delta$. Therefore $\lim_{\delta \to 0} \mathcal{H}_\delta^k(S)$ exists in the extended real number system. The limit is called the k-*dimensional Hausdorff measure* of S and is denoted by $\mathcal{H}^k(S)$. The function $\mathcal{H}^k$ is an outer measure.

Exercises for the Reader

IV.1. If $I \subseteq \mathbb{R}^N$ is a line segment, then $\mathcal{H}^1(I)$ is the usual Euclidean length of I. Also, $\mathcal{H}^0(I) = \infty$ and $\mathcal{H}^k(I) = 0$ for all $k > 1$.

IV.2. If $S \subseteq \mathbb{R}^N$ is Borel, then $\mathcal{H}^N(S) = \mathcal{L}^N(S)$, where $\mathcal{L}^N$ is Lebesgue N-dimensional measure.

IV.3. If $S \subseteq \mathbb{R}^N$ is a discrete set, then $\mathcal{H}^0(S)$ is the number of elements of S.

IV.4. Define $M_\alpha = \Gamma(1/2)^\alpha / \Gamma(1 + \alpha/2)$, and $\alpha > 0$ (note that this is consistent with the preceding definition of M_k). Then define H^α for any $\alpha > 0$ by using the Hausdorff construction. Let S be a subset of $\mathbb{R}^N$. Set $\alpha_0 = \sup\{\alpha > 0 : \mathcal{H}^\alpha(S) = \infty\}$. Also compute $\alpha_1 = \inf\{\alpha > 0 : \mathcal{H}^\alpha(S) = 0\}$. Then $\alpha_0 \leq \alpha_1$. This number is called the *Hausdorff dimension* of S. What is the Hausdorff dimension of the Cantor ternary set? What is the Hausdorff dimension of a regularly imbedded, k-dimensional, C^1 manifold in $\mathbb{R}^N$? In fact (see [FOL, p. 325], [FE1]), any rectifiable set $S \subseteq \mathbb{R}^N$ has the property that $\alpha_0 = \alpha_1$.

The measure $\mathcal{H}^{N-1}$ gives one reasonable definition of $d\sigma$ on $\partial\Omega$ when $\Omega \subseteq \mathbb{R}^N$ has C^1 boundary. Now let us give another. If $S \subseteq \mathbb{R}^N$ is closed and $x \in \mathbb{R}^N$, let $\text{dist}(x, S) = \inf\{|x - s| : s \in S\}$. Then $\text{dist}(x, S)$ is finite, and there is a (not necessarily unique) $s_0 \in S$ with $|s_0 - s| = \text{dist}(x, S)$. (*Exercise:* Prove these assertions.) Suppose that $M \subseteq \mathbb{R}^N$ is a regularly imbedded C^1 manifold of dimension $k < N$ (see [HIR]). Let $E \subseteq M$ be compact and, for $\epsilon > 0$, set $E_\epsilon = \{x \in \mathbb{R}^N : \text{dist}(x, E) < \epsilon\}$. Define

$$\sigma_k(E) = \limsup_{\epsilon \to 0^+} \frac{\mathcal{L}^N(E_\epsilon)}{M_{N-k}\epsilon^{N-k}},$$

where $\mathcal{L}^N$ is Lebesgue volume measure on $\mathbb{R}^N$. It can, in fact, be shown that "limsup" may be replaced by "lim." The resulting set-function σ_k is an outer measure. It can be proved that $\mathcal{H}^k(E) = \sigma_k(E)$. The measure σ_k may be extended to more general subsets of M by the usual exhaustion procedures.

Our third definition of area measure is as follows. Let $M \subseteq \mathbb{R}^N$ be a regularly imbedded C^1 submanifold of dimension $k < N$. Let $p \in M$, and let (ψ, U) be a coordinate chart for $M \subseteq \mathbb{R}^N$, as in the definition of "regularly imbedded submanifold." We use the notation $J_\mathbb{R} G$ to denote the Jacobian matrix of the mapping G.

When $E \subseteq U \cap M$ is compact, define

$$m_k(E) = \int_{\psi(E)} \text{vol}\langle J_{\mathbb{R}} \psi^{-1}(x) e_1, \ldots, J_{\mathbb{R}} \psi^{-1}(x) e_k \rangle d\mathcal{L}^k(x).$$

Here e_j is the jth unit coordinate vector, and the integrand is simply the k-dimensional volume of the k-parallelipiped determined by the vectors $J_{\mathbb{R}} \psi^{-1}(x) e_j$, $j = 1, \ldots, k$. We know from calculus [SPI] that this gives a definition of surface area on compact sets $E \subseteq U \cap M$ that coincides with the preceding definitions. The new definition may be extended to all of M with a partition of unity, and to more general sets E by inner regularity.

Finally, we mention that a k-dimensional, C^1 manifold M may be given (locally) in *parametrized form*. That is, for $P \in M$ there is a neighborhood $P \in U_P \subseteq \mathbb{R}^N$ and we are given functions $\phi_1, \ldots, \phi_N$ defined on an open set $W_P \subseteq \mathbb{R}^k$ such that the mapping

$$\Phi = (\phi_1, \ldots, \phi_N) : W_P \to U_P \cap M$$

is C^1, one-to-one, and onto, and the Jacobian of this mapping has rank k at each point of W_P. In this circumstance, we define

$$\tau_k(U_P \cap M) = \int_{x \in W_P} |M_x| \, d\mathcal{L}^k(x),$$

where M_x is defined to be the standard k-dimensional volume of the image of the unit cube in $\mathbb{R}^k$ under the linear mapping $J_{\mathbb{R}} \Phi(x)$. [The object M_x can be defined rather naturally using the language of differential forms—see [SPI]. The definition we have given has some intuitive appeal.]

On a k-dimensional regularly imbedded C^1 submanifold of $\mathbb{R}^N$, we have

$$\mathcal{H}^k = \sigma_k = m_k = \tau_k.$$

Appendix V
Green's Theorem

Here we record the standard form of Green's theorem that is used in harmonic analysis. A derivation of this particular formula from Stokes's theorem appears in [KRA4, Section 1.3]; that reference also contains applications to the theory of harmonic functions. See also [BAK], [KRP].

Theorem: *Let $\Omega \subseteq \mathbb{R}^N$ be a domain with C^2 boundary. Let $d\sigma$ denote area $((N-1)$-dimensional Hausdorff) measure on $\partial\Omega$—see Appendix IV. Let v be the unit outward normal vector field on $\partial\Omega$. Then, for any functions $u, v \in C^2(\bar{\Omega})$, we have*

$$\int_{\partial\Omega} \left(u \frac{\partial v}{\partial \nu} - v \frac{\partial v}{\partial \nu} \right) d\sigma = \int_{\Omega} (u\Delta v - v\Delta u)\, dV.$$

Appendix VI
The Banach-Alaoglu Theorem

Let X be a Banach space and X^* its dual. Assume for the moment that X is separable. For ϕ_j, ϕ elements of X^*, we say that $\phi_j \to \phi$ in the weak-$*$ topology if, for each $x \in X$, $\phi_j(x) \to \phi(x)$. Thus weak-$*$ convergence is pointwise convergence for linear functionals. It induces the weakest topology on X^* under which the point evaluation functionals are continuous.

Theorem: *Assume that the Banach space X is separable. Let $\{\phi_j\}$ be a bounded sequence in X^*. Then there is a subsequence $\{\phi_{j_k}\}$ that converges in the weak-$*$ topology.*

Informally, we often cite the Banach-Alaoglu theorem by saying that "the unit ball in the dual of a Banach space is weak-$*$ compact."

Note that the unit *sphere* in the dual of a (infinite-dimensional) Banach space is *never* weak-∗ compact (e.g., take an orthonormal sequence in Hilbert space).

In case the Banach space X is not separable, then we cannot specify the weak-∗ topology using sequences; we must instead specify a sub-basis for the topology. See [RUD3] for the details.

Appendix VII
Expressing an Integral in Terms of the Distribution Function

Let f be a measurable function on $\mathbb{R}^N$. For $\alpha > 0$ we set

$$\mu_f(\alpha) = m\{x \in \mathbb{R}^N : |f(x)| > \alpha\}.$$

Then μ_f is the *distribution function* of f. Integrals of the form $\int_{\mathbb{R}^N} \phi(|f|)\, dV$ may be expressed in terms of integrals of $\mu_f(\alpha)$. For example,

$$\int_{\mathbb{R}^N} |f(x)|^p\, dx = \int_0^\infty \mu_f(\alpha) p\alpha^{p-1} d\alpha.$$

For a proof, test the assertion on f the characteristic function of an interval, and then use standard approximation arguments.

Appendix VIII
The Stone-Weierstrass Theorem

Let X be a compact metric space. Let $C(X)$ be the algebra of continuous functions on X equipped with the supremum norm. Let $\mathcal{A} \subseteq C(X)$ be an algebra of continuous functions that contains the constant function 1. [Here an *algebra* is a vector space with a notion of multiplication—see [LAN].] The Stone-Weierstrass theorem gives conditions under which $\mathcal{A}$ is dense in $C(X)$.

Theorem: *Assume that, for every* $x, y \in X$ *with* $x \neq y$, *there is an* $f \in \mathcal{A}$ *such that* $f(x) \neq f(y)$ *(we say that* $\mathcal{A}$ *separates points). [In case the functions in* $\mathcal{A}$ *are complex-valued,* $\overline{f} \in \mathcal{A}$ *whenever* $f \in \mathcal{A}$.*] Then* $\mathcal{A}$ *is dense in* $C(X)$.

Appendix IX
Landau's $\mathcal{O}$ and o Notation

Sometimes a good piece of notation is as important as a theorem. Landau's notation illustrates this point.

Let f be a function defined on a neighborhood in $\mathbb{R}^N$ of a point P. We say that f *is* $\mathcal{O}(1)$ *near* P if

$$|f(x)| \leq C,$$

for some constant C, in that neighborhood of P. Now let g be another function defined on the same neighborhood of P. Writing this last inequality somewhat pedantically as

$$|f(x)| \leq C \cdot 1,$$

we are led to define f to be $\mathcal{O}(g)$ near P if

$$\left|f(x)\right| \leq C \cdot \left|g(x)\right|,$$

for some constant C, in that neighborhood of P.

If instead, for any $\epsilon > 0$, there is a $\delta > 0$ such that

$$\left|f(x)\right| \leq \epsilon \cdot \left|g(x)\right|$$

when $0 < |x - P| < \delta$, then we say that $f = o(g)$ at P.

We will see ample illustration of the Landau notation in our study of Fourier analysis.

Table of Notation

Section	Notation	Meaning
0.2	$\mathbb{R}$	real numbers
0.2	$(a, b), [a, b)(a, b], [a, b]$	intervals
0.2	χ_S	characteristic function
0.2	$s(x)$	simple function
0.2	$\mathbb{R}^N$	Euclidean space
0.2	$\int f(x)\,dx$	Lebesgue integral
0.2	f^+	positive part of function
0.2	f^-	negative part of function
0.2	LMCT	Lebesgue monotone convergence theorem
0.2	LDCT	Lebesgue dominated convergence theorem
0.2	FL	Fatou's lemma
0.2	f_x , f^y	slice functions
0.2	L^p	Lebesgue space
0.2, 0.3	$\mathbb{T}$	circle group
0.2	$\mathcal{A}$	a σ-algebra

Section	Notation	Meaning
0.2	δ_0	Dirac measure
0.2	μ	a measure
0.3	X	a Banach space
0.3	$\| \ \|$	a norm
0.3	$\| \ \|_p$, $\| \ \|_{L^p}$	L^p norm
0.3	$\| \ \|_\infty$, $\| \ \|_{L^\infty}$	essential sup norm
0.3	L^p_{loc}	local L^p space
0.3	$\| \ \|_X$	Banach space norm
0.3	$L : X \to Y$	linear mapping of spaces
0.3	$\| \ \|_{\mathrm{op}}$	operator norm
0.3	X^*	dual space of X
0.3	L^*	adjoint of L
0.4	$\langle \cdot , \cdot \rangle$	inner product
0.4	$\| \ \|$	Hilbert space norm
0.4	H	Hilbert space
0.4	$\perp$	perpendicular or orthogonal
0.4	$\mapsto$	maps to
0.4	$K^\perp$	set of vectors perpendicular to (annihilator of) K
0.4	$A + B$	linear sum of spaces
0.4	ℓ^2	little Lebesgue space
0.5	FAPI	Functional Analysis Principle I
0.5	FAPII	Functional Analysis Principle II
1.1	$\partial_t u$	partial derivative in t
1.1	$\partial_x u$	partial derivative in x

Section	Notation	Meaning
1.1	$f^{(k)}$	k^{th} derivative
1.1	S	unit circle in the complex plane
1.1	τ_g	translation by g
1.1	$\mathbb{C}^*$	the multiplicative group $\mathbb{C} \setminus \{0\}$
1.1	$\widehat{f}(j)$	j^{th} Fourier coefficient
1.1	$Sf \sim \sum\limits_{j=-\infty}^{\infty} \widehat{f}(j)e^{ijt}$	formal Fourier expansion
1.1, App I	C_c^k, C_c^∞	compactly supported, smooth functions
1.2	$S_N f$	partial sum of Fourier series
1.2	$D_N(s)$	Dirichlet kernel
1.2	$f * g$	convolution of functions
1.2, App IX	$\mathcal{O}(t)$	Landau's notation
1.3	$f(x+)$	right limit of f at x
1.3	$f(x-)$	left limit of f at x
1.3	Λ	sequence of scalars
1.3	$\mathcal{M}_\Lambda$	multiplier operator induced by Λ
1.3	$\sigma_N f$	Cesàro mean of f
1.3	K_N	Fejér kernel
1.3	P_r	Poisson kernel for the disc
1.3	$P_r f$	Abel sum of f
1.4	$\{k_N\}$	a family of kernels
1.4	$A \approx B$	A is comparable in size to B
1.5	Mf	Hardy-Littlewood maximal function

Section	Notation	Meaning
1.5	$P^* f$	maximal Poisson integral of f
1.6	$\mathbf{d}(f, g)$	metric induced by L^2 topology
1.6, App VIII	$C(\mathbb{T})$	set of continuous functions on $\mathbb{T}$
1.6	$\operatorname{sgn} x$	signum function
1.6	P.V. $\int dx$	Cauchy principal value
1.6	Hf	Hilbert transform of f
1.6	$\widetilde{H} f$	modified Hilbert transform of f
1.6	Jf	complexified Hilbert transform of f
1.6	e_j	the jth exponential function
1.6	$f * K$	convolution of a function with a kernel
2.1	$t \cdot \xi$	inner product on Euclidean space
2.1	$\widehat{f}(\xi)$	Fourier transform
2.1	$(\ )^{\widehat{}}$	Fourier transform
2.1	$\gg$	much greater than
2.1	C_0	continuous functions that vanish at ∞
2.2	$O(N)$	orthogonal group on $\mathbb{R}^N$
2.2	ρ	a rotation
2.2	α_δ , α^δ	dilation operators
2.2	τ_a	translation operator
2.2	$\widetilde{f}$	odd reflection of f

Section	Notation	Meaning
2.3	G_ϵ	Gaussian summability kernels
2.3	$\check{g}(x)$	inverse Fourier transform
2.3	$\mathcal{F}$	Fourier transform
2.4	z	complex variable
2.4	Var f	variance of f
2.4	$[A, B]$	commutator of A and B
3.1	$\widetilde{S}_N f$	alternative method of partial summation
3.1	$\mathbb{T}^2$	torus group
3.1	$|(j, k)|$	modulus of an index
3.1	$S_R^{\text{sph}} f$	spherical partial sum
3.1	S_M^{sq}	square partial sum
3.1	$S_{(m,n)}^{\text{rect}}$	rectangular partial sum
3.1	$S_R^{\text{poly}, P}$	polygonal partial sum
3.3	$B(0, 1)$	unit ball in Euclidean space
3.3	$\mathcal{M}_B$	multiplier operator for the ball
3.3	$\mathcal{M}_m$	multiplier operator induced by m
3.3	Q_N	fundamental region in $\mathbb{T}^N$
3.3	$\eta(x)$	$(2\pi)^{-N/2} e^{-|x|^2/2}$
3.3	$\eta_{\sqrt{\epsilon}}(x)$	$(2\pi)^{-N/2} \epsilon^{N/2} e^{-\epsilon|x|^2/2}$
3.3	Λ	fundamental lattice in $\mathbb{R}^N$
3.3	$\widetilde{S} f$	periodized multiplier operator induced by S
3.4	Q	unit cube in $\mathbb{R}^N$

Section	Notation	Meaning
3.4	Q_R	dilate of unit cube
3.4	$H\phi$	Hilbert transform of ϕ
3.4	$E_{\mathbf{v}}$	half space determined by vector $\mathbf{v}$
3.5	$\triangle ABC$	triangle with vertices A, B, C
3.5	T_A'	triangular sprouts
3.5	$T_{jk\ell}$	sprouted triangles
3.5	Σ_k	unit sphere in $\mathbb{R}^k$
3.5	$\mathcal{T}_j$	half space multiplier operator
3.5	$\widetilde{R}$	adjunct rectangles to R
3.5	$\widetilde{\widetilde{R}}$	subadjunct rectangles to R
3.5	$|E|$	Lebesgue measure of the set E
4.1	$f_r(e^{i\theta})$	circular slice of f
4.1	$F_{j,r}$	j^{th} Fourier coefficient of f_r
4.1	$\mathcal{H}_k$	spherical harmonics of degree k
4.1	ρ_ϕ	rotation through angle ϕ
4.1	$\alpha = (\alpha_1, \ldots, \alpha_N)$	multi-index
4.1	$|\alpha|$	modulus of a multi-index
4.1	x^α	multi-index product notation
4.1	$\frac{\partial^\alpha}{\partial x^\alpha}$	multi-index derivative notation
4.1	$\alpha!$	multi-index factorial notation

Section	Notation	Meaning
4.1	$\delta_{\alpha\beta}$	Kronecker delta
4.1	$\mathcal{P}_k$	space of homogeneous polynomials of degree k
4.1	d_k	dimension of $\mathcal{P}_k$
4.1	$P(D)$	differential polynomial
4.1	$\langle P, Q \rangle$	inner product on homogeneous polynomials
4.1	$\triangle$	Laplace operator
4.1	$\ker \phi_k$	kernel of ϕ_k
4.1	$\operatorname{im} \phi_k$	image of ϕ_k
4.1	$\mathcal{A}_k$	space of solid spherical harmonics of degree k
4.1	$Y^{(k)}$	a spherical harmonic of degree k
4.1	$\partial/\partial v$	unit outward normal derivative
4.1	$\sigma,\ d\sigma$	rotationally invariant surface measure
4.1	a_k	dimension of the space $\mathcal{H}_k$
4.2	$e_{x'}$	point evaluation functional
4.2	$Z_{x'}^{(k)}$	zonal harmonic
4.2	$P(x, t')$	Poisson kernel for the ball
4.2	$P_k^\lambda(t)$	Gegenbauer polynomial
4.2	J_k	k^{th} Bessel function
5.1	$\mathcal{D}^{2j}$	$(2j)^{\text{th}}$-order differentiation operator
5.1	k_{2j}	kernel associated to $\mathcal{D}^{2j}$

Section	Notation	Meaning
5.1	$\mathcal{I}^{2j}$	$(2j)^{\text{th}}$-order integration operator
5.1	$\mathcal{I}^{\beta}$	β^{th}-order fractional integration operator
5.2	$K(x)$	Calderón-Zygmund kernel
5.2	$\Omega(x)/|x|^N$	Calderón-Zygmund kernel
5.2	T_K	Calderón-Zygmund singular integral operator
5.2	$\Omega(x)$	numerator of Calderón-Zygmund kernel
5.2, 5.8	BMO	functions of bounded mean oscillation
5.4	D	unit disc in the complex plane
5.4	$H^p(D)$	p^{th}-order Hardy space on D
5.4	$H^\infty(D)$	space of bounded holomorphic functions
5.4	f^*	boundary limit function of f
5.4	$\mathbf{h}^p(D)$	p^{th}-power integrable harmonic functions on D
5.4	$\mathbf{h}^\infty(D)$	bounded harmonic functions on D
5.4	$B_a(z)$	Blaschke factor
5.4	$B(z)$	Blaschke product
5.4	B_K	partial Blaschke product
5.4	B^*	boundary function of a Blaschke product B

Section	Notation	Meaning
5.5	v	conjugate harmonic function to the harmonic function u
5.5	$\widetilde{\phi}$	boundary function for v
5.5	$\widetilde{k}(t) = \cot(t/2)$	Hilbert transform kernel on disc
5.5	H_{Re}^1	real-variable Hardy space of order 1
5.5	$K_j(x)$	j^{th} Riesz kernel
5.5	R_j	j^{th} Riesz transform
5.5	$\\| \\ \\|_{H_{\text{Re}}^1}$	real-variable Hardy space norm
5.5	$P_y(x)$	Poisson kernel for the upper half-space
5.6	Mf	Hardy-Littlewood maximal operator
5.6	f^*	maximal operator associated to the kernel ϕ_0
5.7	$a(t)$	atom
5.7	H_{Re}^p	real-variable Hardy space of order p
5.7	$\\| \\ \\|_*$	BMO norm
5.7	Q	cube in $\mathbb{R}^N$
6.1	$\rho(x, y)$	quasi-metric
6.1	$B(x, r)$	ball determined by a quasi-metric
6.1	(X, ρ, μ)	space of homogeneous type
6.1	C_1, C_2	constants for space of homogeneous type

Section	Notation	Meaning
6.1	$\rho(z, w)$	nonisotropic distance on the unit ball of $\mathbb{C}^n$
6.1	$\mathcal{M}_x$	presentation of the Heisenberg group
6.2	$R_\alpha(x)$	kernel of fractional integral operator
6.2	E_y , E^x	partial distribution functions
6.2	$Mf(x)$	Hardy-Littlewood maximal function
6.2	$M^* f(x)$	modified Hardy-Littlewood maximal function
6.2	$\mathcal{O} = \bigcup_j B(o_j, r_j)$	Whitney decomposition
6.2	$\delta_\mathcal{O}$	distance to the complement of $\mathcal{O}$
6.2	$f = g + \sum_j h_j$	Calderón-Zygmund decomposition
6.2	T_k	integral operator induced by k
6.2	K_N	truncated kernel
6.3	$a(x)$	atom
6.3	H_{Re}^1	real-variable or atomic Hardy space
6.3	$T(\mathbf{1})$ theorem	on L^2-boundedness of integral operators
6.3	T^t	transpose of the operator T
6.3	$T(b)$ theorem	generalization of the $T(\mathbf{1})$ theorem

Section	Notation	Meaning
6.3	$\lambda(j)$	coefficients in the Cotlar-Knapp-Stein theorem
6.3	$\phi^{x,\epsilon}$	translate and dilate of ϕ
7.3	MRA	Multi-Resolution Analysis
7.3	ϕ	wavelet scaling function
7.3	ψ	wavelet
7.3	V_j	subspaces in an MRA decomposition
7.3	W_j	wavelet subspaces
7.3	**MRA$_1$**	scaling axiom for an MRA
7.3	**MRA$_2$**	inclusion axiom for an MRA
7.3	**MRA$_3$**	density axiom for an MRA
7.3	**MRA$_4$**	maximality axiom for an MRA
7.3	**MRA$_5$**	basis axiom for an MRA
7.4	f_N	approximation to the Dirac delta
7.4	$\widetilde{D}_N(t)$	$\sin Nt/\pi t$
7.5	$m(\xi)$	low-pass filter
App II	$\rho_{\alpha,\beta}$	Schwartz semi-norms
App II	$\mathcal{S}$	the Schwartz space
App IV	$\mathcal{H}_\delta^k$	approximate Hausdorff measure
App IV	$\mathcal{H}^k$	Hausdorff measure
App IV	$\alpha_0,\ \alpha_1$	lower and upper Hausdorff dimensions
App IV	E_ϵ	ϵ-thickening of the set E

Section	Notation	Meaning
App IV	σ_k	k-dimensional measure defined by thickening
App IV	m_k	k-dimensional measure defined by pullback
App IV	τ_k	k-dimensional measure by parametrization
App VI	weak-$*$ topology	pointwise topology on a dual space
App VII	μ_f	distribution function of f
App VIII	$C(X)$	algebra of continuous functions on X
App VIII	$\mathcal{A}$	subalgebra of $C(X)$
App IX	$\mathcal{O}, o$	Landau's notation

Bibliography

[AHL] L. Ahlfors, *Complex Analysis*, 3rd Ed., McGraw-Hill, 1979.

[ASH1] J. M. Ash, Multiple trigonometric series, in *Studies in Harmonic Analysis*, J. M. Ash, ed., Mathematical Association of America, Washington, D.C., 1976, 76–96.

[ASH2] J. M. Ash, A new proof of uniqueness for trigonometric series, *Proc. AMS* 107(1989), 409–410.

[ASH3] J. M. Ash, Uniqueness of representation by trigonometric series, *Am. Math. Monthly* 96(1989), 873–885.

[BAK] J. A. Baker, Integration over spheres and the divergence theorem for balls, *Am. Math. Monthly* 104(1997), 36–47.

[BEC] W. Beckner, Inequalities in Fourier analysis, *Annals of Math.* 102(1975), 159–182.

[BES] A. Besicovitch, On Kakeya's problem and a similar one, *Math. Z.* 27(1928), 312–320.

[BOU] J. Bourgain, Spherical summation and uniqueness of multiple trigonometric series, *Internat. Math. Res. Notices* 3(1996), 93–107.

[CALZ] A. P. Calderón and A. Zygmund, On the existence of certain singular integrals, *Acta Math.* 88(1952), 85–139.

[CAR] L. Carleson, On convergence and growth of partial sums of Fourier series, *Acta Math.* 116(1966), 135–157.

[CHKS1] D. C. Chang, S. G. Krantz, and E. M. Stein, Hardy Spaces and Elliptic Boundary Value Problems, Proceedings of a Conference in Honor of Walter Rudin, *Contemporary Math.* 137, Am. Math. Society, 1992.

[CHKS2] D. C. Chang, S. G. Krantz, and E. M. Stein, H^p theory on a smooth domain in $\mathbb{R}^N$ and applications to partial differential equations, *Jour. Funct. Anal.* 114(1993), 286–347.

[CHO] G. Choquet, *Lectures on Analysis*, in three volumes, edited by J. Marsden, T. Lance, and S. Gelbart, Benjamin, New York, 1969.

[CHR] F. M. Christ, *Singular Integrals*, CBMS, Am. Math. Society, Providence, 1990.

[COI] R. R. Coifman, A real variable characterization of H^p, *Studia Math.* 51(1974), 269–274.

[COIW1] R. R. Coifman and G. Weiss, *Analyse Harmonique Non-Commutative sur Certains Espaces Homogenes*, Springer Lecture Notes vol. 242, Springer-Verlag, Berlin, 1971.

[COIW2] R. R. Coifman and G. Weiss, Extensions of Hardy spaces and their use in analysis, *Bull. AMS* 83(1977), 569–645.

[CON] B. Connes, Sur les coefficients des séries trigonometriques convergents sphériquement, *Comptes Rendus Acad. Sci. Paris* Ser. A 283(1976), 159–161.

[COO] R. Cooke, A Cantor-Lebesgue theorem in two dimensions, *Proc. Am. Math. Soc.* 30(1971), 547–550.

[COR1] A. Cordoba, The Kakeya maximal function and the spherical summation multipliers, *Am. Jour. Math.* 99(1977), 1–22.

[COR2] A. Cordoba, The multiplier problem for the polygon, *Annals of Math.* 105(1977), 581–588.

[CORF1] A. Cordoba and R. Fefferman, On the equivalence between the boundedness of certain classes of maximal and multiplier operators in Fourier analysis, *Proc. Nat. Acad. Sci. USA* 74(1977), 423–425.

[CORF2] A. Cordoba and R. Fefferman, On differentiation of integrals, *Proc. Nat. Acad. Sci. USA* 74(1977), 221–2213.

[COT] M. Cotlar, A combinatorial inequality and its applications to L^2-spaces, *Revista Math. Cuyana* 1(1955), 41–55.

[CUN] F. Cunningham, The Kakeya problem for simply connected and for star-shaped sets, *Am. Math. Monthly* 78(1971), 114–129.

[DAU] I. Daubechies, *Ten Lectures on Wavelets*, Society for Industrial and Applied Mathematics, Philadelphia, 1992.

[DAVJ] G. David and J. L. Journé, A boundedness criterion for generalized Calderón-Zygmund operators, *Ann. Math* 120(1984), 371–397.

[DAY] C. M. Davis and Yang-Chun Chang, *Lectures on Bochner-Riesz Means*, London Mathematical Society Lecture Note Series 114, Cambridge University Press, Cambridge–New York, 1987.

[DUS] N. Dunford and J. Schwartz, *Linear Operators*, Wiley-Interscience, New York, 1958, 1963, 1971.

[FE1] H. Federer, *Geometric Measure Theory*, Springer-Verlag, Berlin and New York, 1969.

[FEF1] C. Fefferman, Inequalities for strongly singular convolution operators, *Acta Math.* 124(1970), 9–36.

[FEF2] C. Fefferman, The multiplier problem for the ball, *Annals of Math.* 94(1971), 330–336.

[FEF3] C. Fefferman, The uncertainty principle, *Bull. AMS*(2) 9(1983), 129–206.

[FEF4] C. Fefferman, Pointwise convergence of Fourier series, *Annals of Math.* 98(1973), 551–571.

[FEF5] C. Fefferman, On the convergence of Fourier series, *Bull. AMS* 77(1971), 744–745.

[FEF6] C. Fefferman, On the divergence of Fourier series, *Bull. AMS* 77(1971), 191–195.

[FES] C. Fefferman and E. M. Stein, H^p spaces of several variables, *Acta Math.* 129(1972), 137–193.

[FOL1] G. B. Folland, *Real Analysis: Modern Techniques and their Applications*, John Wiley and Sons, New York, 1984.

[FOL2] G. B. Folland, *A Course in Abstract Harmonic Analysis*, CRC Press, Boca Raton, Florida, 1995.

[FOS] G. B. Folland and A. Sitram, The uncertainty principle: a mathematical survey, *J. Fourier Anal. and Appl.* 3(1997), 207–238.

[FOST] G. B. Folland and E. M. Stein, Estimates for the $\bar{\partial}_b$ complex and analysis on the Heisenberg group, *Comm. Pure Appl. Math.* 27(1974), 429–522.

[FOU] J. Fourier, *The Analytical Theory of Heat*, Dover, New York, 1955.

[GAC] J. Garcia-Cuerva, Weighted H^p spaces, thesis, Washington University in St. Louis, 1975.

[GAR] J. Garnett, *Bounded Analytic Functions*, Academic Press, New York, 1981.

[GRK] R. E. Greene and S. G. Krantz, *Function Theory of One Complex Variable*, John Wiley and Sons, New York, 1997.

[GUZ] M. de Guzman, *Differentiation of Integrals in $\mathbb{R}^n$*, Springer Lecture Notes, Springer-Verlag, Berlin and New York, 1975.

[HAN] Y. S. Han, Triebel-Lizorkin spaces on spaces of homogeneous type, *Studia Math.* 108(1994), 247–273.

[HEI] M. Heins, *Complex Function Theory*, Academic Press, New York, 1968.

[HEL] S. Helgason, *Differential Geometry and Symmetric Spaces*, Academic Press, New York, 1962.

[HERG] E. Hernandez and G. Weiss, *A First Course on Wavelets*, CRC Press, Boca Raton, Florida, 1996.

[HERS] E. Hewitt and K. Ross, *Abstract Harmonic Analysis*, vol. 1, Springer-Verlag, Berlin, 1963.

[HIR] M. Hirsch, *Differential Topology*, Springer-Verlag, Berlin and New York, 1976.

[HOF] K. Hoffman, *Banach Spaces of Analytic Functions*, Prentice-Hall, Englewood Cliffs, New Jersey, 1962.

[HOR] L. Hörmander, L^p estimates for (pluri-)subharmonic functions, *Math. Scand.* 20(1967), 65–78.

[HUN] R. Hunt, On the convergence of Fourier series, *1968 Orthogonal Expansions and Their Continuous Analogues* (Proc. Conf., Edwardsville, Illinois, 1967), pp. 235–255. Southern Illinois University Press, Carbondale, Illinois.

[HUT] R. Hunt and M. Taibleson, Almost everywhere convergence of Fourier series on the ring of integers of a local field, *SIAM J. Math. Anal.* 2(1971), 607–625.

[JON] F. John and L. Nirenberg, On functions of bounded mean oscillation, *Comm. Pure and Appl. Math.* 14(1961), 415–426.

[JOUR] J. L. Journé, Calderón–Zygmund operators, pseudodifferential operators and the Cauchy integral of Calderón, *Springer Lecture Notes* 994, Springer-Verlag, Berlin, 1983.

[KAT] Y. Katznelson, *Introduction to Harmonic Analysis*, Wiley, New York, 1968.

[KNS1] A. Knapp and E. M. Stein, Singular integrals and the principal series, *Proc. Nat. Acad. Sci. USA* 63(1969), 281–284.

[KNS2] A. Knapp and E. M. Stein, Intertwining operators for semisimple Lie groups, *Annals of Math.* 93(1971), 489–578.

[KON] J. J. Kohn and L. Nirenberg, On the algebra of pseudodifferential operators, *Comm. Pure and Appl. Math.* 18(1965), 269–305.

[KONS] S. V. Konyagin, On divergence of trigonometric Fourier series over cubes, *Acta Sci. Math.* (Szeged) 61(1995), 305–329.

[KOR] T. J. Körner, *Fourier Analysis*, 2nd ed., Cambridge University Press, Cambridge, 1989.

[KRA1] S. Krantz, *Real Analysis and Foundations*, CRC Press, Boca Raton, Florida, 1992.

[KRA2] S. Krantz, Fractional integration on Hardy spaces, *Studia Math.* 73(1982), 87–94.

[KRA3] S. Krantz, *Partial Differential Equations and Complex Analysis,* CRC Press, Boca Raton, Florida, 1992.

[KRA4] S. Krantz, *Function Theory of Several Complex Variables,* 2nd Ed., Wadsworth Publishing, Belmont, California, 1992.

[KRL] S. Krantz and S.-Y. Li, Hardy classes, integral operators, and duality on spaces of homogeneous type, preprint.

[KRP] S. Krantz and H. Parks, *The Geometry of Domains in Space,* manuscript.

[LAN] S. Lang, *Algebra,* 2nd ed., Addison-Wesley, Reading, Massachusetts, 1984.

[LAN] R. E. Langer, *Fourier Series: The Genesis and Evolution of a Theory,* Herbert Ellsworth Slaught Memorial Paper I, *Am. Math. Monthly* 54(1947).

[LAT] R. H. Latter, A characterization of $H^p(\mathbb{R}^N)$ in terms of atoms, *Studia Math.* 62(1978), 93–101.

[LUZ] N. Luzin, The evolution of "Function," Part I, Abe Shenitzer, ed., *Am. Math. Monthly* 105(1998), 59–67.

[MAS] R. A. Macias and C. Segovia, Lipschitz functions on spaces of homogeneous type, *Adv. Math.* 33(1979), 257–270.

[MEY1] Y. Meyer, *Wavelets and Operators,* Translated from the 1990 French original by D. H. Salinger, Cambridge Studies in Advanced Mathematics 37, Cambridge University Press, Cambridge, 1992.

[MEY2] Y. Meyer, *Wavelets. Algorithms and Applications,* translated from the original French and with a foreword by Robert D. Ryan, SIAM, Philadelphia, 1993.

[MEY3] Y. Meyer, Wavelets and operators, in *Analysis at Urbana 1,* E. R. Berkson, N. T. Peck, and J. Uhl eds., London Math. Society Lecture Note Series, Vol. 137, Cambridge University Press, Cambridge, 1989.

[MOS1] J. Moser, A new proof of de Giorgi's theorem concerning the regularity problem for elliptic differential equations, *Comm. Pure and Appl. Math.* 13(1960), 457–468.

[MOS2] J. Moser, On Harnack's theorem for elliptic differential equations, *Comm. Pure and Appl. Math.* 14(1961), 577–591.

[NSW] A. Nagel, E. M. Stein, and S. Wainger, Balls and metrics defined by vector fields. I. Basic properties, *Acta Math.* 155(1985), 103–147.

[RUD1] W. Rudin, *Principles of Mathematical Analysis,* 3rd ed., McGraw-Hill, New York, 1976.

[RUD2] W. Rudin, *Real and Complex Analysis*, McGraw-Hill, New York, 1966.

[RUD3] W. Rudin, *Functional Analysis*, McGraw-Hill, New York, 1973.

[SCH] L. Schwartz, *Théorie des Distributions*, Hermann, Paris, 1957.

[SJO1] P. Sjölin, An inequality of Paley and convergence a.e. of Walsh–Fourier series, *Ark. Math.* 7(1969), 551–570.

[SJO2] P. Sjölin, On the convergence almost everywhere of certain singular integrals and multiple Fourier series, *Ark. Math.* 9(1971), 65–90.

[SMI] K. T. Smith, A generalization of an inequality of Hardy and Littlewood, *Can. Jour. Math.* 8(1956), 157–170.

[SPI] M. Spivak, *Calculus on Manifolds*, Benjamin, New York, 1965.

[STE1] E. M. Stein, *Singular Integrals and Differentiability Properties of Functions*, Princeton University Press, Princeton, 1970.

[STE2] E. M. Stein, *Harmonic Analysis: Real-Variable Methods, Orthogonality and Oscillatory Integrals*, Princeton University Press, Princeton, 1993.

[STE3] E. M. Stein, Harmonic analysis on $\mathbb{R}^N$, in *Studies in Harmonic Analysis*, J. M. Ash ed., Math. Assn. of America, Washington, D.C., 1976, pp. 97–135.

[STG1] E. M. Stein and G. Weiss, *Introduction to Fourier Analysis on Euclidean Spaces*, Princeton University Press, Princeton, 1971.

[STG2] E. M. Stein and G. Weiss, On the theory of harmonic functions of several variables, *Acta Math.* 103(1960), 25–62.

[STR1] R. Strichartz, *A Guide to Distribution Theory and Fourier Transforms*, CRC Press, Boca Raton, Florida, 1993.

[STR2] R. Strichartz, Para-differential operators—another step forward for the method of Fourier, *Notices of the A.M.S.* 29(1982), 402–406.

[STR3] R. Strichartz, How to make wavelets, *Am. Math. Monthly* 100(1993), 539–556.

[TAW] M. Taibleson and G. Weiss, The molecular characterization of certain Hardy spaces, *Representation Theorems for Hardy Spaces*, pp. 67–149, *Astérisque* 77, Soc. Math. France, Paris, 1980.

[TEV] N. Tevzadze, On the convergence of double Fourier series of quadratic summable functions, *Soobšč. Akad. Nauk. Gruzin. SSR.* 5(1970), 277–279.

[UCH] Uchiyama, A maximal function characterization of H^p on the space of homogeneous type, *Trans. AMS* 262(1980), 579–592.

[WAL] J. S. Walker, Fourier analysis and wavelet analysis, *Notices* of the AMS 44(1997), 658–670.

[WAT] G. N. Watson, *A Treatise on the Theory of Bessel Functions*, Cambridge University Press, Cambridge, 1922.

[WHI] H. Whitney, *Geometric Integration Theory*, Princeton University Press, Princeton, 1957.

[WIC] M. V. Wickerhauser, *Adapted Wavelet Analysis from Theory to Software*, A. K. Peters, Wellesley, 1994.

[WIE] N. Wiener, *The Fourier Integral and Certain of its Applications*, Dover, New York, 1958.

[ZYG] A. Zygmund, *Trigonometric Series*, Cambridge University Press, Cambridge, 1959.

Index